OPERAMONDOBOOK

AF720345

Nazzareno Luigi Todarello

The Particle of God

Unveiling the Secrets of the Universe

LATORRE PRESS

THE PARTICLE OF GOD: Unveiling the Secrets of the Universe
Nazzareno Luigi Todarello
First Latorre Press edition 2025

Cover design: Nazzareno Luigi Todarello
Interior design: Latorre Press
Publisher: Latorre Press, Viale della Rimembranza 23/2
15067 Novi Ligure AL Italy
www.latorrepress.com
www.latorre-editore.it

ISBN (paperback): 9791282487085

I. THE INFINITELY LARGE AND THE INFINITELY SMALL

It's incredibly difficult to grasp the scale of things that are extremely large or extremely small. Our imagination, our capacity to envision, is tied to our experiences—the experiences of our lives. What we don't see directly eludes us. Or rather, we can imagine what we don't see, certainly, but only if its characteristics aren't too far removed from the things we know from direct experience. Our imagination, in reality, has short legs. Who can imagine, in a single glance, the route of their car if it exceeds a few dozen kilometers? We can only do so by zooming out on the navigation screen. This is a first point: there are technological tools today that fuel our imagination and help it move into unknown territories. Up to a certain point, however. If we learn, for example, that in a glass of water, say eighteen grams, there are about six hundred thousand billion billion water molecules—a six followed by twenty-three zeros—it is absolutely impossible for us to form an idea of the size of a water molecule. We cannot imagine something so small. We are irremediably conditioned by our own dimensions and the dimensions of the things that make up our world. And yet, the water molecule is not the smallest object that exists. It is made up of two hydrogen atoms and one oxygen atom. And in turn, each atom is made up of much smaller particles: protons, neutrons, and electrons. And the electrons are very far from the nucleus formed by protons and neutrons. So much so that one must imagine the atom as almost completely empty, with the nucleus and electrons very far apart. And each proton is made up of quarks, which are also smaller than our imagination can comprehend. And here, for now, we stop. For now, because it is not excluded that in the future new technological means will allow us to find even smaller particles. For now, quarks are the extreme horizon of the enormously small. In the sense that this is where certainties end. Then there are only hypotheses. Hypotheses that wait to be proven to become scientific truth. String theory, for example, conceived from an intuition of the Italian physicist Gabriele Veneziano, hypothesizes that everything is in reality, in its ultimate essence, a set of tiny vibrating rings and that the particles we know are each a manifestation of a particular set of vibrations. It must be said that the Italian translation of "strings theory" should more properly be "theory of strings," violin strings, or of another stringed instrument. If we were to actually identify a string, to see it and measure it, the entire universe would appear to us as an immense concert, an almost endless orchestra in perpetual vibration. Then the ancient proposition proclaiming "the harmony of the spheres" would find an unexpected confirmation. We'll see. Physics is not a world of fantastic things. Imagination plays a big part in the intuition of new theories, but then every new theory must be confirmed to become "true," that is, "measurable." There is no doubt that imagination played a decisive role in Einstein's mind when, while he was a clerk at the patent office in Bern, he imagined the theory of relativity. He then formalized his intuition in mathematical formulas, but his theory became reality when experiments confirmed it with certain, incontrovertible measurements. Here: physics deals with measurable things. (And in fact, I must say, the title of this chapter, with that adverb "infinitely" repeated twice, is not very "physical"). And over time, it has learned to measure incredibly small things. Just as it has learned to measure incredibly large things. This is the other aspect of modern physics that leaves our imagination astonished and powerless. Can we imagine something that travels at three hundred thousand kilometers per second? And can we imagine the distance that thing covers in an hour, in a year? Far beyond the route of our car on the navigation screen! And yet our home, the Earth, is twenty-six thousand light-years away from the black hole at the center of the galaxy of which it is a part. Light travels. For us, it travels in an astonishing way. But in the immensities of space, it seems to move like a snail.

Even staying within our galaxy to go from one place to another takes years, thousands of years.

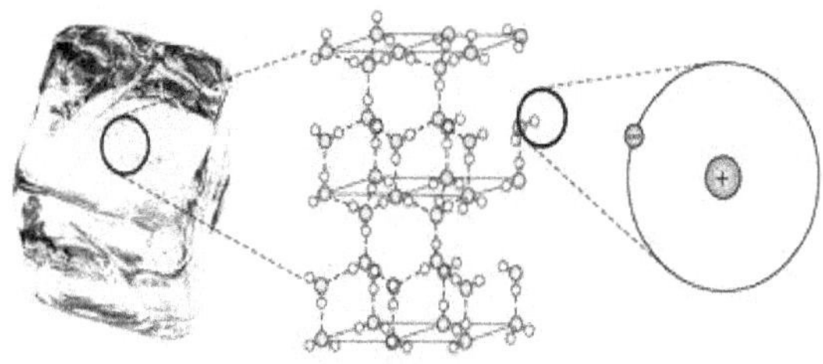

Figure 1

A 2.7-centimeter ice cube weighs about 18 grams. It consists of a lattice structure containing a little over six hundred thousand billion billion water molecules, H2O. Each oxygen atom contains eight protons and eight neutrons, while each hydrogen atom contains only one proton. The ice cube therefore contains approximately 10.8 million billion billion protons and neutrons. (from Jim Baggott, The Higgs Boson, 2013).

Here we are now, between the immensely small and the immensely large. It almost makes you think that we are in the right place. Maybe so. Or maybe not. We don't know yet. But we do know that the immensely small and the immensely large are related. Not only because, obviously, the immensely large is made up of an immense number of immensely small things. But also, and above all, because by studying the immensely small, we can understand phenomena of the immensely large. With the particle accelerator at CERN, in fact, we investigate the origins of the universe. In a twenty-seven-kilometer tunnel, protons are accelerated to a speed close to that of light. From one side and the other so that at a certain point they collide, generating something very similar to the Big Bang. The protons, colliding, break into pieces. A small inferno of heat is unleashed. Particles are born. The newborns are photographed and studied at length. There are confirmations and, often, surprises.

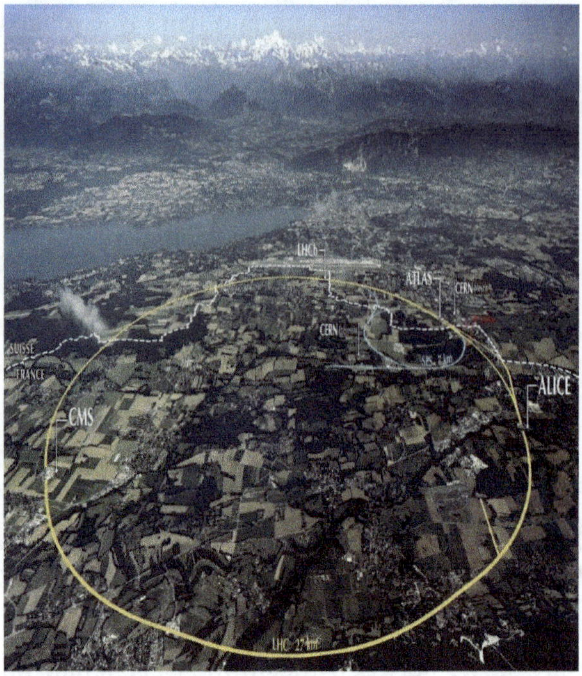

Figure 2

CERN (Conseil européen pour la recherche nucléaire), is the world's largest laboratory for particle physics. It is located at the border between Switzerland and France, on the western outskirts of Geneva. The convention establishing CERN was signed on September 29, 1954, by 12 member states. Today, it has 2400 employees, hosts visiting scientists, and collaborates with 608 universities and 113 countries worldwide. The aerial photograph shows the ring of the LHC (Large Hadron Collider). In reality, the tunnel is a hundred meters underground.

Figure 3

The LHC tunnel. When it's turned on, protons race through the tube at a speed close to that of light. Enormous magnets cooled with liquid helium guide the protons inside the tube. These magnets are the coldest place in the universe because they are at a temperature below 3 Kelvin, colder than the temperature of outer space. Hadron is any particle composed of three quarks.

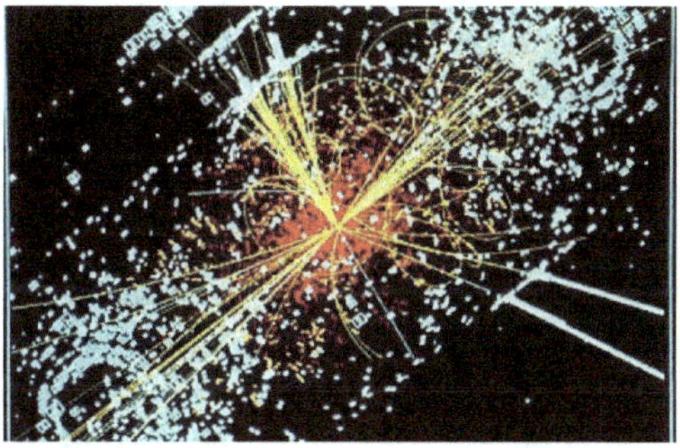

Figure 4

When hadrons collide, an enormous amount of energy is released, giving birth to countless particles with extremely short lifetimes. Particle detectors photograph them at a rate of millions of frames per second.

II. THE INDIVISIBLE

It all began a very long time ago. Exactly two thousand and five hundred years ago. When Greek philosophers asked themselves questions that we still ask ourselves today. Mainly: what is the world made of? And how is it made? Perhaps before them, no one asked themselves these questions? Certainly not. Everyone, for a long time, had asked themselves similar questions. But the difference is that the Greek philosophers are the first to seek answers through reasoning and not by relying on fantasy. It is the birth, precisely, of philosophy, of rational thought. Philosophy takes the place of myth, that is, of the fantastic story of the origins of the world. Be careful, however, we must not be too harsh with myth. We risk considering our ancestors to be foolish. And that was not the case. Myths perfectly fulfilled their function. They gave a convincing answer to the fear of the unknown, to the desire to know. And they served above all to keep together human beings belonging to the same social group. Around the fire, in the evening, the stories of the elders filled minds with wonder and consolidated the sense of belonging. But things changed quickly in classical Greece. Cities took over from the countryside. Citizens became worldly. They trade, they know other peoples, other myths. They need other truths. Myths lose strength. Their historical function is exhausted. Now organized, concrete, logical reasoning responds to the new type of man, to his questions. The questions themselves are asked differently. The pre-Socratics pose the problem of the origins of the cosmos and give answers that, although still clothed in mysterious poetry, already know something new. Socrates, of Athens, investigates the logical mechanisms of knowledge, unmasking any certainty based on nothing and any presumption. Plato, a student of Socrates, organizes the logical space of the West, clearly posing the essential problems. In short, he prepares the playing field on which the human reason of the West will play its games. Aristotle, a student of Plato, observes, reasons, catalogs, explains. He wraps everything in a web of reasoning of excellent quality. His observations are often brilliant: "The closer a thing is to its essence, the closer it is to the truth." We can say that contemporary physics does exactly this: it goes deep into things to know their ultimate reality, their essence. Aristotle gives a logical answer to every question. But the philosopher who interests us most now is Democritus, of whom we have no writing. We know the titles of his books, about fifty, but the texts have been lost forever. Perhaps it was the first Christians who destroyed them. We don't know, but it's possible. The early Christians were sectarian and fanatical like the Taliban of today. It is possible, probable, that they destroyed the books of the philosopher who denied the eternity of the soul. We know of Democritus what other authors tell us about him and his thought. Democritus is the great theorist of atomism. That is, he is the one who first hypothesized in a coherent way that everything that exists is made up of atoms. The thing is amazing. Democritus had no tools to investigate matter. He knew what his senses showed him. And his senses showed him only what they could show him. What they also show us. Macroscopic forms. We see microscopic forms today. We see them because we have a microscope. Democritus did not have a microscope. How did he understand that everything is made of atoms? How do scientists do it: by observing and reasoning. He observed and saw around him that everything and every being changes continuously. And yet, he observed, everything and every being, although changing continuously, remains itself. How can this happen? Democritus understood that the transformation did not concern so much the macroscopic forms, which do change but in an imperceptible way. The transformation had to take place at a lower level, not in the whole but in the parts. He therefore imagined that every form is made up of invisible parts, so small that they cannot be further subdivided. These very small parts aggregate according to visible forms that evolve coherently for a certain period. Like the shapes that flocks of birds make in the sky. Or like the waves of the dunes that change shape slowly but continuously because the grains, always the same, move continuously driven by the wind. Ever-changing forms, because the individual elements never stand still. Forms, however, that maintain their unity in change. He called those parts "atoms" which means "uncuttable", not further divisible. Only after more than twenty centuries were human beings able to actually see atoms. And only then did it become clear that the atom was not the smallest part of nature. Because there were nuclei, protons, electrons, quarks. The philosophical principle to which Democritus had adhered, and before him his master (and first "atomist" philosopher) Leucippus, was that beyond a certain limit matter cannot be divided because otherwise you would reach a point where there is nothing more, where matter loses consistency. The principle of the conservation of matter was at the basis of ancient atomism: matter exists anyway, it cannot dissolve into fragments too similar to nothingness. Aristotle did not share this way of understanding things, who thought instead that matter was continuous and therefore infinitely divisible. Who was right? Aristotle was wrong, because today we know that matter is not infinitely divisible. When you get to subatomic particles you can no longer divide. Today we know that matter is very close to the consistency of a vacuum. Today we know that the void is not empty, because there are fields. Today we know that matter is an expression of energy and that therefore every matter can lose its "material consistency" and transform into energy. As difficult as it is to believe, this is precisely what modern physics has demonstrated. Today we are convinced that mass is not an intrinsic or "primary" property of the fundamental constituents of nature. In practice, mass does not exist. It is only the consequence of the energy of the interactions involving elementary particles. In short, physicists have continued to divide, and in the end they have found nothing. And today we ask ourselves a new question: how does energy transform into matter? In other words: how does a particle of energy acquire mass, that is, begin to weigh?

III. WHY DO THINGS WEIGH?

It seems like an absurd question. And in fact, not even the most intelligent Greek philosophers ever asked it clearly. It's so natural that things weigh. They didn't ask the question, not even the medieval monk philosophers. Those philosophers whom the modern world unjustly slights. Unjustly, because they were as intelligent as the Greeks and made excellent use of their brains, while all around was a frightening forest and minds numbed by fear and need. Yet they suspected something, both the ancients and the monks. The monks especially had an open account with matter. The account was open because of the body of Christ. What do we really swallow, they asked themselves, when we swallow the consecrated host? The body of Christ, is the answer. Yes, but how does it happen? The simple faithful were content with the word of the Church. The monk philosopher wanted to understand. What is matter if it can transform from a blade of wheat into a human body? And from fermented grape juice into blood? It is "transubstantiation", the mystical process through which a substance is transferred from one body to another. This is the point: to distinguish between matter and substance. The philosopher Egidio Romano, who lived in the 13th century, called after death "Doctor fundatissimus" and "Theologorum princeps", was the one who most consciously distinguished between form/matter and substance/content. The material form is determined by the dimensions. It is what we see of the object, its sensible appearance. The substance is the real content of the object, which escapes the gaze. Egidio Romano gave the example of ice. We see a block of ice, we appreciate its shape and dimensions. We can measure them. But when the heat melts the ice we realize that the previous dimensions were not the substance of the object. The substance is water, which can take different forms, expanding and contracting. It can also evaporate and transform into a cloud and then condense again into water and solidify, changing volume again, into ice. The shape changes, the volume, but the "quantity of matter", the substance, remains unchanged. Egidio Romano's distinction between volume and mass is still valid. Egidio Romano lived between 1243/47 and 1316. He studied in Paris. He was exempted from teaching for his ideas. Then he became General of the Order of Saint Augustine and tutor to Philip the Fair, etc. At that time science was clothed in theology. Perhaps that is why we modern inhabitants of the West, who have all decided together that only what is material is real, know nothing of that thought. It seems to us the infancy of knowledge. And yet little by little we are discovering that everything that is material, in its "substance", is very little material. We should be less arrogant in our materialistic certainties. Things weigh. Why? All things are made of particles. Some particles do not weigh, that is, they have no mass. And they are the particles that have the privilege of traveling at the speed of light. Indeed, to tell the truth, those particles "are" the light and travel at a crazy speed, as we have seen and as everyone now knows. The photon is in fact the particle/light, that is, it is a small amount of energy that runs bouncing back and forth illuminating and heating. Then there are particles that weigh, that is, they have a mass. This prevents them from moving at the speed of photons. This has been established, without a shadow of a doubt, by the glorious mind of Einstein. And a century of experiments has confirmed it. A proton can go, if pushed adequately, at a speed close to that of light (closer and closer, if the push continues to increase), but without ever reaching it. Because, Einstein said, to reach it the push would have to be infinite. This is because as it approaches the speed of light, the proton increases in mass. More and more the closer it gets to that fateful speed. If it reached it (and to reach it it would take an infinite push of energy) the proton would have infinite mass. Which is not possible. Doubly impossible. So at a certain point any push, no matter how enormous its energy, no longer has any effect on acceleration. Or rather, it has a smaller and smaller effect. Because the mass to be accelerated is too great

and the more you increase the energy with which you push it, the more the mass increases. Giving energy to a body means increasing its mass. If, hypothetically, energy continued to be input forever, the speed of the body would increase forever, but always for smaller and smaller fractions of acceleration, without ever being able to reach the speed of light, that of massless particles. In short, the body would behave like Achilles in Zeno's famous paradox. The proton is made of quarks. And quarks have mass. And other particles have mass. Why them and not the photon? And, even stranger, why don't particles that have mass all have the same mass? Why do some weigh more and others less? The neutron and the proton, for example, which even if they are embraced that no one can separate to form the nucleus of any atom, do not have the same mass. The neutron weighs a little more. Little, but more. And the top quark, which is the smallest part of creation to date, weighs more or less like a gold atom, which is made of 79 protons, 79 neutrons and 79 electrons.
(I have used, trivially, the word weight to indicate mass. We earthlings usually do so. To be precise, weight is the force with which one body attracts another body. It is called gravitation. It is not an intrinsic quality of the body. If the mass of the body that attracts it changes, the attracted body changes weight. Mass, on the other hand, is intrinsic to the body and is the quantity of matter that makes up the body itself. And, to be even more precise, it is the sum of the quantity of matter and the quantity of energy that make up a body).

IV. THE STANDARD MODEL

When it was understood, at the end of the 19th century, that the atom was not indivisible, the study of subatomic particles began. During the 20th century, a definitive list was compiled, at least for now. The collection of known particles and the forces that make them interact form the so-called "Standard Model". The Standard Model has given great satisfaction to physicists worldwide, although it suffers from significant gaps. Let's briefly examine it.

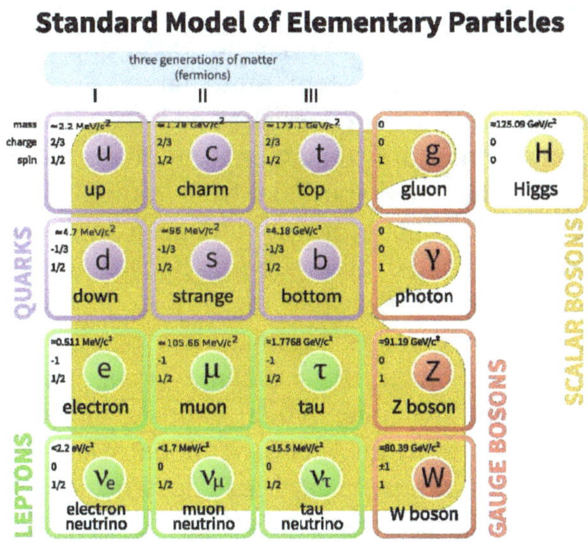

Figure 5

The Standard Model is the result of decades of hypotheses and experiments. In its simplicity, it represents the culmination of one of the most fruitful periods of scientific research, from the late 19th century to the 1970s. There is also the Higgs boson, whose presence is indispensable for the coherence of the whole. However, its existence was not experimentally confirmed until 2012. The Standard Model does not contain any information about gravity or dark energy and dark matter.

In Figure 5, we see twelve matter particles: six types of quarks and six types of leptons, or light particles. Quarks and leptons, collectively, are known as "fermions" in honor of the Italian physicist Enrico Fermi. Then, there are four types of force particles, called bosons. The name "boson" is also a tribute to a great physicist, in this case, the Indian physicist Satyendranath Bose. Finally, there's the Higgs boson, a "scalar" boson that we'll discuss later. All of these are considered "elementary" particles, meaning they are indivisible. Two force particles are massless: the gluon and the photon, which are the only two particles without weight. The other three, the Z boson, the W- boson, and the W+ boson, do have mass. The Standard Model only includes particles that have been experimentally observed. For instance, the "graviton," a hypothetical particle that would transmit the force of gravity, has not been detected yet. The number of particles can seem quite large. Some physicists believe that this number is excessive and hypothesize that there must be something more fundamental. Perhaps there exists an even more "elementary" particle from which all others are composed. This idea is not new: we've seen similar reductions in complexity with the atom, the nucleus, and protons and neutrons. It's therefore possible that this could happen again. String theory, mentioned earlier, was developed to address this possibility. Indeed, the ultimate dream of many physicists is to find "the particle," the fundamental building block of everything. For now, however, the Standard Model works remarkably well and explains almost everything. Let's now delve into the characteristics of these various particles.

V. QUARKS, QUARKS, QUARKS

Figure 5 shows six types of quarks. As we've mentioned, quarks combine in groups of three to form protons and neutrons. Thus, they reside at the heart of the atom, within the nucleus.

Before 1964, protons and neutrons were believed to be elementary particles. In 1964, Murray Gell-Mann and George Zweig proposed the existence of smaller particles. Only with this hypothesis could they explain the peculiar behavior of protons and neutrons. By 1968, quarks were experimentally observed, convincing even the most skeptical scientists of their existence. The name "quark" was coined by Murray Gell-Mann, inspired by a nonsensical play on words from James Joyce's "Finnegans Wake," one of the most incomprehensible books ever written. However, there was a precursor to this idea. In 1963, Robert Serber of Columbia University began exploring combinations of three unknown fundamental particles to create the two octets of the Eightfold Way (Figure 42).

When Gell-Mann visited Columbia in March of that year for a series of lectures, Serber sought his opinion on this idea. Their conversation took place during a lunch break at the Columbia faculty club. Serber suggested, "We can take three pieces and construct protons and neutrons, and with pieces and anti-pieces, we could make mesons." Gell-Mann responded dismissively, "If you want to get a proton and a neutron with charges of +1 and 0, the particles would have to have fractional charges (–1/3, 2/3, things like that)."

In the ensuing discussion, Gell-Mann referred to Serber's new particles as "quorks," a meaningless word, mirroring Serber's seemingly nonsensical idea. Serber thought it was a mispronunciation of the English word "quirk," suggesting a strange quirk of nature. However, Gell-Mann realized that this was exactly the right approach. There had to be three particles within protons and neutrons. He wrote down the equations and

assigned the strange name from Joyce to these new particles, reminiscent of the offhand remarks made at Columbia.
Quarks are never found in isolation. No one has ever observed a lone quark. They prefer to exist in groups of three, permanently confined within protons and neutrons. Despite numerous attempts, no one has been able to split them apart. A proton contains two up quarks and one down quark. Physicists, being a rather imaginative bunch, have attributed colors and flavors to quarks. Thus, we have green, red, and blue quarks. Additionally, there are "charm," "strange," "top," and "bottom" quarks. These fanciful names, which have no direct connection to our sensory experience (a "green" quark is not actually green), are used to describe the properties of each type of quark. The situation is quite complex. To simplify, the color of a quark can be thought of as a kind of electric charge, but while electric charges are either positive or negative, quarks have three colors. This is why there are three different color quarks in protons and neutrons. In this way, the three "charges" cancel each other out, and the "baryon" (the family that includes protons and neutrons) is "white," or neutral and stable.
The most interesting thing about quarks is that they interact through the so-called "strong force" (also part of the Standard Model, as shown in Figure 5, where the gluon, from the word "glue," is the carrier of the strong force). The strong force is what binds the nucleus of an atom together. We know that when a nucleus splits, an enormous amount of energy is released. As a result of the strong interaction, quarks can change color. Within the nucleus, gluons are constantly at work, causing quarks to change color. However, stability is maintained because the sum of the colors always remains neutral. There is always one green, one red, and one blue quark, even if the individual quarks are constantly changing. It's like a coordinated dance. This dance is made even more complex by the fact that the "flavors" of quarks can also change: for example, an up quark can transform into a down quark by absorbing or emitting a W boson. This process occurs thanks to the weak interaction, one of the four fundamental forces.

It's also important to know that quarks do not all have the same mass. Some are heavier than others. The heaviest quarks do not exist in nature; they can only be observed during high-energy collisions. It is hypothesized that they were more abundant in the hot, dense universe immediately after the Big Bang.
Quarks are fascinating particles, and there is still much we don't know about them. When quarks are produced in particle accelerators, for example, detectors observe jets of color-neutral particles instead of isolated quarks, as one might expect from fragmentation.

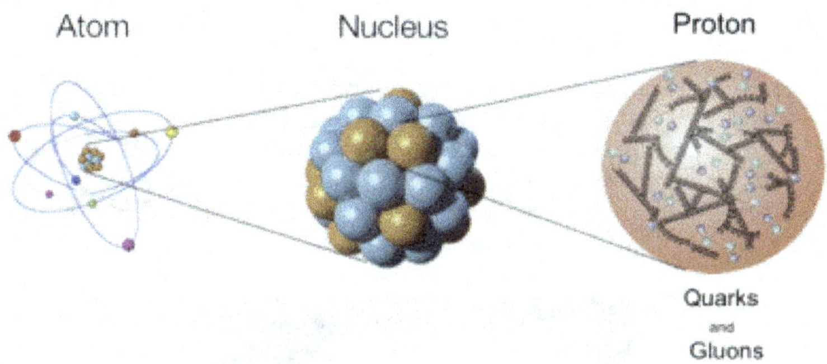

Figure 6

Atom, nucleus, protons, quarks, and gluons. The quantity of protons, neutrons, and electrons in an atom is the same. Electrons have a negative charge, protons have a positive charge, and neutrons have no charge. Therefore, the atom as a whole has a neutral charge. If an atom loses or gains an electron, it becomes an "ion," which is an atom with a charge, negative (if it gains an electron) or positive (if it loses an electron). This process is called "ionization."

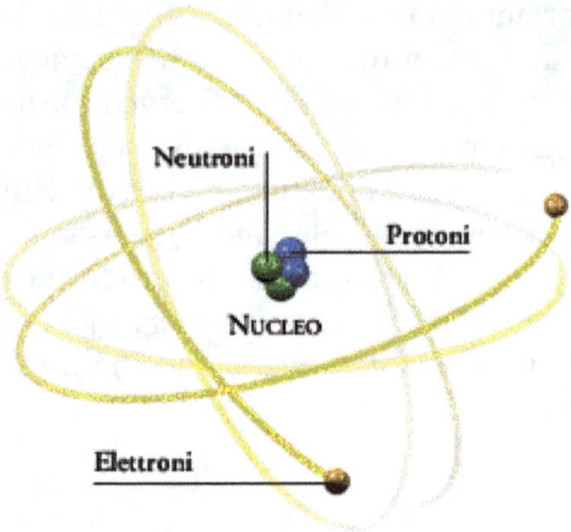

Figure 7

A helium atom is comprised of two electrons, two protons, and two neutrons. The protons and neutrons are confined within the nucleus. Protons and neutrons are composed of quarks of various types. It is important to note that the diagram is not drawn to scale. The immense void separating the electrons from the nucleus cannot be accurately depicted.

Figure 8

A proton consists of a bound state of three quarks: two up quarks and one down quark. The quarks are confined within the proton by the strong interaction, mediated by gluons. The color charge of quarks, which comes in three varieties, is a crucial property. To form a colorless baryon such as a proton, the quarks must have different colors.

VI. LEPTONS

The word "lepton" comes from the Greek and means "light". Leptons are, in fact, almost entirely massless. Electrons, which orbit the nucleus and complete the marvel of nature that is the atom, are leptons. Neutrinos are also leptons, a most mysterious family of particles. Neutrinos do not interact with anything. They pass through any body with aristocratic ease, not deigning to make contact with anything. The sun produces billions upon billions of billions, etc., of neutrinos per second. Billions of neutrinos pass through the Earth. Billions of neutrinos pass through our bodies. Nothing changes as they pass. This led to the long-held belief that the neutrino was completely massless. However, in 1998, Japanese scientists at the Super-Kamiokande observatory showed that the neutrino does have mass, albeit a very small one: one hundred thousand to one million times smaller than the mass of the electron. This also revealed that they do, in fact, undergo some very small interactions. It has been calculated that a wall of lead one light-year thick would only be able to stop half of the neutrinos that entered it. The other half would pass through as if nothing were there. In any case, even a very little interaction is more than nothing. So even neutrinos, despite their near-zero mass and zero charge, cannot escape some form of interaction. As a result, until it was thought that neutrinos had no mass at all, they were considered to be involved in the matter of dark matter. Once it was discovered that they have some mass, their involvement seemed much less likely.
But why is the neutrino called a "neutrino"? Perhaps it is a diminutive of "neutron"? That's exactly right. The nickname was coined by Edoardo Amaldi during a conversation with Enrico Fermi at the Physics Institute in Via Panisperna in Rome, as a playful diminutive of "neutron," also a neutral but

much more massive particle. The term was then adopted by Fermi at a conference in Paris in July 1932 and at the Solvay Conference of 1933, where it was also used by Wolfgang Pauli, who had postulated its existence in 1930. Thus baptized by the two greats, the particle definitively took that name. But in reality, the neutrino had not yet been born. There was the name, but not the child, which was born only twenty years later, in 1956, when physicists Clyde Cowan and Fred Reines "saw" it in an experiment conducted at the Savannah River fission reactor.
The true protagonist of the "lepton" group is the famous electron. It is 0.06% of the atom but determines its chemical properties. It is a reference point. The masses of other particles are calculated in relation to the mass of the electron. And the electric charge as well. In fact, we speak of electronvolts. The electronvolt (the electric charge of an electron multiplied by a volt) is the unit of measurement for electric charges in the atomic and subatomic realm.
In conclusion, quarks and electrons, bound together in the atom, constitute all the stable known matter in the universe. The other particles of the Standard Model exist, but it is not known why. They seem to serve no purpose at all.

VII. RUTHERFORD'S GOLD

The concept of the atom has been discussed since antiquity. However, a scientifically rigorous description of these entities emerged only at the beginning of the 20th century. In 1902, Joseph John Thomson proposed the first significant atomic model. According to this theory, the atom was conceived as a sphere uniformly charged positively, within which negatively charged electrons were embedded. Thomson, at that time, was unaware of the existence of protons and neutrons. His atomic model represented the atom as a dense solid, with electrons embedded within it, similar to raisins in a plum pudding. For this reason, this model is historically known as the "plum pudding model".

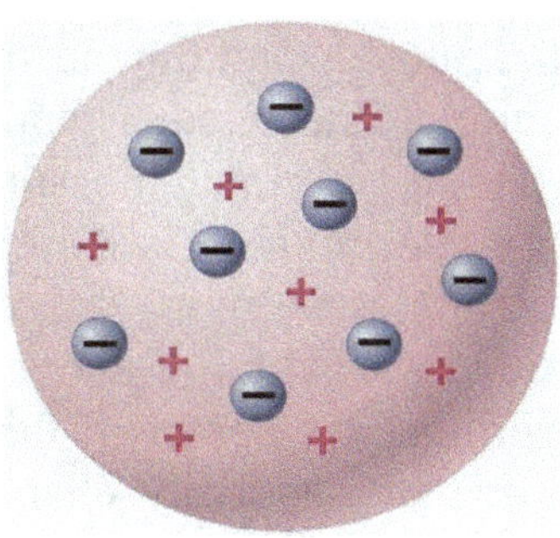

Figure 9

Thomson's atomic model posits that the entire mass of the atom carries a positive charge. Negatively charged electrons are embedded within this positively charged mass.

Shortly after Thomson's "plum pudding" model, it was replaced by a new atomic model. In 1910, two of Rutherford's students (Geiger and Marsden) conducted a decisive experiment. As often happens, the theory that scientists set out to confirm with their experiment ended up being destroyed by the experiment itself. In fact, in physics, we speak of both "verification" and "falsification." Geiger and Marsden, when they organized the 1910 experiment, intended to verify/falsify Thomson's atomic theory. They discovered things that that theory did not predict. They bombarded a very thin sheet of gold with alpha particles. Behind the gold foil was a screen that detected the alpha particles that passed through the foil. Alpha particles, which are particles with low penetrating power, should bounce off or be deflected by the gold atoms, and the screen should detect this. The astonishing result was that only a few particles, one percent, bounced off or were deflected. All the others passed through the gold foil, going straight to the screen. The atom, therefore, is not full and compact. Ernest Rutherford realized that the mass of the atom is not diffused, but concentrated at the center, the nucleus. The electrons revolve very far away. In the middle is a vacuum. It is the "planetary model": the nucleus at the center like the sun and electrons orbiting like planets. The distance between the nucleus and the electrons is ten thousand to one hundred thousand times greater than the diameter of the nucleus itself. These are minimal measurements in absolute terms, but relatively enormous. If we imagine the hydrogen nucleus as big as an apple, we must imagine the electron as a grain of sand orbiting at a radius of about one kilometer. Rutherford also realized that the center of the system is not monolithic but composed of positively charged particles bound together: protons. And he also realized that the total mass of the nucleus cannot be justified by the sum of the masses of the protons that constitute it and that therefore there must be other particles. In 1932, James Chadwick discovered the particles intuited by Rutherford. These are particles with no charge: neutrons.

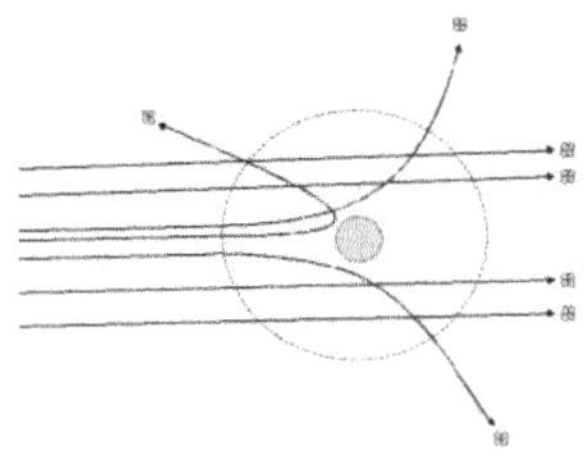

Figure 10

Alpha particles pass unimpeded through the atomic void of gold, with only a minority experiencing deflection or rebound upon encountering the centrally positioned nucleus. This experimental observation was first made by Geiger and Marsden in 1910.

However, the new model presents a significant challenge: what prevents the electron from spiraling into the nucleus? Based on the principles of electromagnetism, an electron orbiting the nucleus should radiate electromagnetic energy, such as light, causing it to lose energy and ultimately collide with the nucleus. If this were true, atomic stability would be impossible, and the universe as we know it could not exist. This apparent paradox suggests that either Rutherford's planetary model or the established theory of electromagnetism is fundamentally flawed.

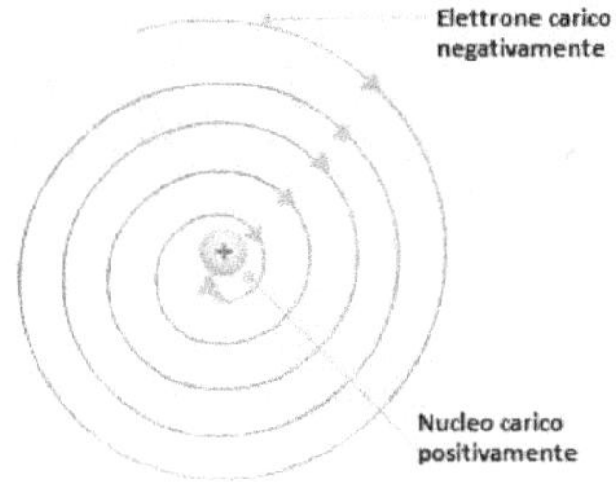

Figure 11

Based on the principles of electromagnetism, an accelerating charged particle, such as an electron, emits electromagnetic radiation. Consequently, an electron orbiting a nucleus would continuously lose energy, causing it to spiral inward and eventually collide with the nucleus in an extremely short timeframe, on the order of less than a millionth of a millionth of a second.

There is another puzzling aspect to this. Experiments involving the emission and absorption of electromagnetic radiation (such as light or heat) reveal a distinctive pattern: each type of atom emits a unique set of wavelengths, corresponding to specific colors, and no others. This implies that all matter, when heated, radiates light, but only at particular wavelengths. Conversely, when exposed to a broad spectrum of radiation, matter selectively absorbs only certain frequencies. The question remains: why do atoms exhibit such specific and selective behaviors regarding the emission and absorption of electromagnetic radiation?

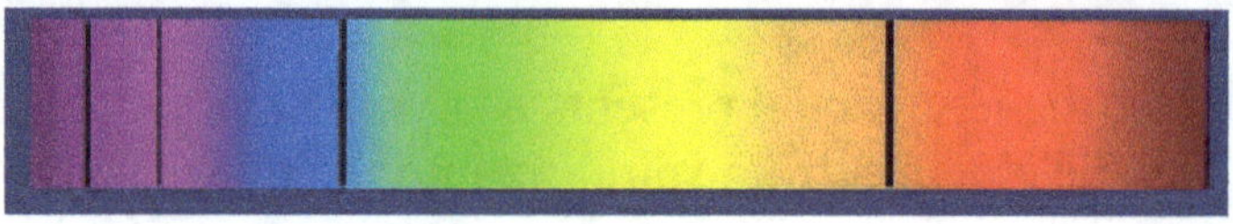

Figure 12

Hydrogen absorption spectrum. The dark lines coincide with the emission frequencies depicted in Figure 13. Frequencies falling outside these lines are not absorbed by the hydrogen atoms but are instead reflected or transmitted, making them visible to the observer.

Figure 13

Hydrogen emission spectrum. The emission spectrum is complementary to the absorption spectrum. Rutherford's atomic model fails to account for the observed emission and absorption spectra of hydrogen and other elements.

The planetary model of the atom is grounded in classical physics, specifically Newton's law of universal gravitation. This law postulates that the force of attraction between two bodies is directly proportional to the product of their masses and inversely proportional to the square of the distance separating them. Given the success of Newtonian mechanics on the macroscopic scale, it was initially assumed that these principles would extend seamlessly to the atomic realm. However, experimental evidence has revealed phenomena that are incompatible with classical mechanics. For instance, if the electron behaved like a planet orbiting a star, it should be capable of occupying any orbit at any distance from the nucleus. Furthermore, heated atoms should emit a continuous spectrum of light, as the excited electrons could occupy any energy level. Yet, atomic spectra exhibit discrete lines, indicating that only specific energy transitions are allowed. Moreover, the classical model fails to explain the stability of the atom, as electrons should spiral into the nucleus due to energy loss through radiation. The persistence of atoms, despite the predictions of classical physics, necessitates a more sophisticated model of the atom.

VIII. THE BOHR ATOM

The early years of the twentieth century were extraordinary for physics. New experiments and new theories followed one another with impressive frequency.

We have seen that Rutherford's atom immediately entered into crisis for two reasons. First: it does not explain why the electron does not fall into the nucleus. Second, it does not explain those colored lines that can be seen in the emission and absorption spectra of bodies. Not even the spectrum of the gas with the simplest atom, hydrogen (Figures 12 and 13).

Three years after the gold foil experiment, in 1913, Niels Bohr, a Danish physicist, solved the problem of spectra with a brilliant hypothesis. The planetary model is valid, Bohr says, but the orbits are not random. The electron runs along precise orbital paths, corresponding to its charge. Each possible orbit is therefore characterized by a "quantized" energy. "Quantized" means that in no case can the electron move randomly. The electron moves in jumps, in "leaps". Each higher orbit is characterized by a charge that is an integer multiple of the charge of the innermost orbit, the "quantum". When the electron is not excited, it does not absorb or emit energy. But when it is stimulated by a particular photon, the electron absorbs the charge and changes orbit, jumping to the higher one. When the excitation ends, the electron returns to its ground state and jumps down, releasing the photon it previously acquired: a photon of light of a defined energy, that is, a light wave of a given color. To understand, it's like the lanes in stadiums: each runner must run inside the lane, they cannot run on the white line that divides them. It's not that there are pre-established lanes around the nucleus. It is the electron itself that traces them, but it can only trace those specific lanes depending on the amount of energy it has to spend: the number two lane with twice the

energy of the number one, the three with triple the one, etc. Hydrogen, when heated, emits energy only at certain frequencies. If exposed to light with all frequencies, it absorbs the same frequencies that it is able to emit. An absorption spectrum (Figure 12) is therefore the negative of the emission spectrum (Figure 13). But why exactly those lines, those colors?
Bohr's answer: electrons are located in well-defined orbits and can rotate indefinitely without absorbing or emitting energy. But when they are hit by a light radiation, they absorb certain wavelengths (Figure 12), jumping to higher orbits: that's why some lines are missing in the spectrum, because they are absorbed by the electrons to jump to higher orbits. When, however, the electrons return to their natural orbit, they emit the energy previously absorbed and the lines of the emission spectrum reappear (Figure 13). In practice, the electron chooses how much energy (i.e., which frequencies, being the energy of the electromagnetic wave its frequency) it can use to make one or more jumps and absorbs it. As we have seen, only charges one, two, three times greater than its own. The other frequencies do not interest it and it sends them back to the sender (and it is the light that we see when an element burns). Once it has absorbed new charge, it moves away from the nucleus by a precise amount, exactly as much as the new charge allows it. In other words, the lengthening of the radius of the electron's orbit is directly proportional to the increase in charge

.

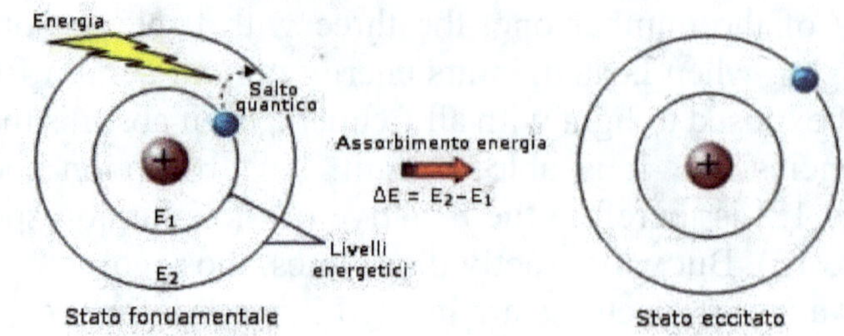

Figure 14

Excited by heat, the electron absorbs energy and transitions to a higher energy level, a process known as a quantum leap. This is the absorption state (as depicted in Figure 12).

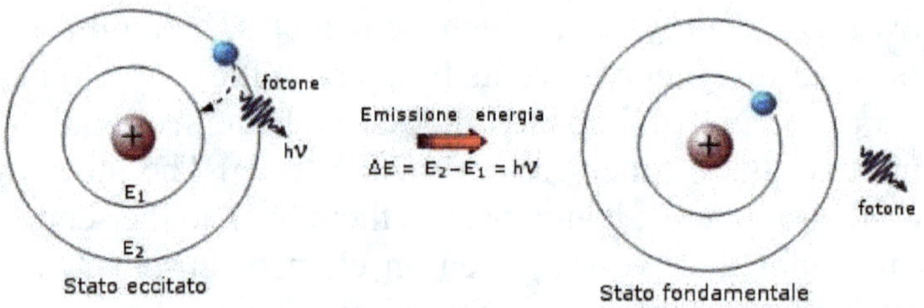

Figure 15

After the energy excitation ceases, the electron relaxes back to its ground state, emitting the photon it had previously absorbed. This is the emission state (as depicted in Figure 13).

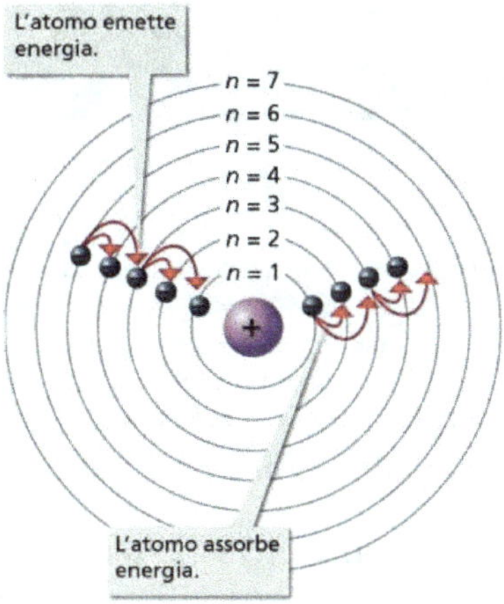

Figure 16

Bohr Model of the Atom. The electron is restricted to specific orbits, each associated with a discrete energy level that is an integer multiple of the ground state energy. Higher energy levels correspond to larger orbits. To transition to a higher energy level, the electron must absorb energy, such as heat or electrical energy. Conversely, when the electron drops to a lower energy level, it emits a photon with a specific frequency, resulting in a colored line in the emission spectrum. The energy of the emitted or absorbed photon is equal to the difference in energy between the initial and final orbits.

Bohr's atomic model was not universally accepted at first. However, it laid the foundation for quantum theory, which would have far-reaching implications. The fact that each element has a unique emission and absorption spectrum is a direct consequence of the arrangement of electrons in different orbits around the nucleus. This unique spectral fingerprint allows us to identify the elements present in any substance, even those located in distant celestial bodies. For instance, by analyzing the emission spectra of stars, astronomers can determine their chemical.

Figure 17

Iron Emission Spectrum. An element's emission spectrum is the characteristic pattern of frequencies of light emitted when its electrons transition from higher to lower energy levels. Given that iron atoms possess 26 electrons, the number of possible electronic transitions within these atoms is significantly greater than that of hydrogen, which has only one electron (as depicted in Figure 13).

Bohr wins the Nobel Prize in 1922, a well-deserved recognition of his genius. Yet, there are still questions that his atom does not answer. First, why doesn't the electron fall into the nucleus? An old and crucial question. Second, why must the orbits be precisely those? It is now experimentally proven that the electron's energy levels are discrete and not continuous. Bohr writes of "the inadequacy of classical electrodynamics to describe the behavior of a system of atomic dimensions." This is an acceptance of an experimental fact, not refutable. However, it is not an explanation. That it is so is now evident to all, but the theoretical context, the final and complete explanation, is missing. What else is there to discover in that marvel of mysteries that the atom is proving to be? So small and so difficult to understand!

Physical science is in crisis in these years. But young geniuses are about to enter the research ring. They are young, less than thirty years old. They are curious. They are not assisted by research institutes. There are not yet the great scientific organizations of today, which provide scientists with impressive financial and organizational resources. These are guided only by curiosity. They want to understand how the matter of which the world is made works. They are Heisenberg, Schrödinger, de Broglie.

These young people have a historic task. They must bring order to a sea of data, accumulated over decades of increasingly precise experiments. A sea of irrefutable data. Irrefutable and incomprehensible.

IX. THE DUKE

The fact is that a colossal mental leap is needed. A great effort of imagination. Bohr's electron is still a ball. A very special ball, of course, running at an insane speed in its small circular world. Jumping from one track to another if someone gives it a shock. But, as soon as it can, it returns to its favorite circle. It has its weight, small but real. It is a material fragment. A small planet. Something is wrong. This ball is strange. Everyone knows it, the scientists of the time, but no one knows what to say. It would take another genius, a genius of the caliber of Rutherford and Bohr.

Maurice de Broglie is a well-known physicist when he participates in a congress of scientists at the Solvay Congress in Brussels. The Solvay, starting in 1911, are an important appointment for physicists. In them, the state of the art on a particular sector of physical research is taken stock of. Maurice de Broglie (a scion of an ancient Piedmontese family who moved to France and a duke by birth) is an experimental physicist, a member in 1924 of the Académie des Sciences, and in 1934 he will be elected to the prestigious Académie Française. Together with him, in Brussels, has arrived his younger brother, Louis de Broglie. He is here as a secretary. No one imagines that it will be him, the younger brother, who will solve the now agonizing problem of the electron and its orbits. In contact with so much intelligence of physicists, young de Broglie decides to dedicate himself body and soul to scientific research. He already has a couple of degrees: "letters and law", and "natural sciences". But now it is the new quantum physics that ignites his imagination. Since he is a nobleman, he shuts himself up in his castle, to study. In 1924, the same year his brother joins the Académie des Sciences, his doctoral thesis is published: "Research on quantum theory". And here is the great find, the brilliant idea:

matter possesses a dual aspect. It is corpuscular, but it is also wave-like. Like the electromagnetic radiation explained by Einstein. It seems impossible. And yet it is so. The electron behaves both as a particle and as a wave. The theory is called the "Dualistic theory of the electron". According to de Broglie, just as electromagnetic radiations have a dual nature, wave-like and corpuscular, so electrons in turn have a dual nature, corpuscular and wave-like. To an electron, as to all moving bodies, a wavelength can be associated. De Broglie describes the incredible with a simple equation:

$$\lambda = m / (h \cdot v)$$

which reads: "the wavelength of the electron lambda (λ) is equal to the mass of the electron itself (m) divided by Planck's constant (h) multiplied by the velocity of the electron (v)."
So the orbits of the electron around the nucleus do not have the circular shape they had in Bohr's atom.

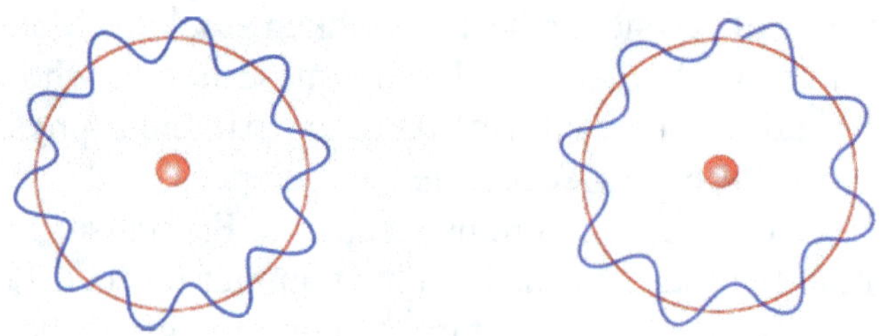

Figure 18

The circular orbits represent Bohr's quantized electron orbits. The sinusoidal waves illustrate de Broglie's wave-like nature of the electron. For stability, the electron's wave must form a standing wave within the orbit (left), meaning the orbit's circumference is an integer multiple of the wavelength. In contrast, if the wave cannot form a complete standing wave (right), the electron's state would be unstable and short-lived.

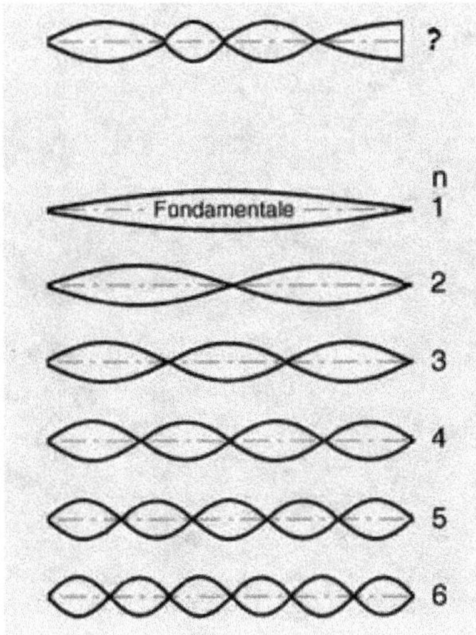

Figure 19

Imagine the electron as a wave circling the nucleus. Waves, like those on a vibrating string, have specific patterns. A piano string vibrates in modes where the ends are always still (nodes). The electron wave must also form a standing wave, with nodes at specific points along the orbit. This means the circumference of the orbit must be an exact multiple of the wavelength. If the wave cannot form a complete standing wave, it will quickly collapse.

The reason electron orbits have specific energy levels is due to the wave-like nature of the electron. The electron's wave must form a complete standing wave around the nucleus, meaning it must fit an exact number of wavelengths within the orbit. If the wave were to be interrupted, the electron would be in an unstable state. As the electron gains energy, its orbit expands to accommodate a larger number of wavelengths. This wave-particle duality prevents the electron from spiraling into the nucleus.

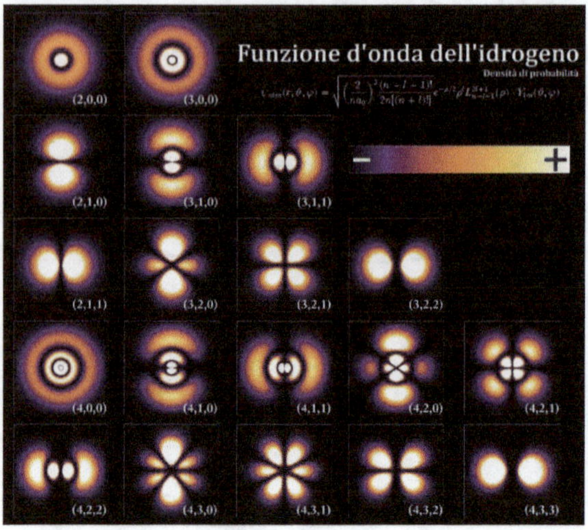

Figure 20

The electron's motion around the nucleus is described by an orbital, which is essentially a probability cloud. This cloud represents the three-dimensional space where the electron is likely to be found. The shape of this cloud depends on the secondary quantum number l. For l = 0, the orbital is a sphere. As l increases to 1, 2, and 3, the orbital takes on more complex shapes with two, four, and eight lobes, respectively. The nucleus of the atom is always located at the center of these lobes

De Broglie's hypothesis is experimentally confirmed in 1927. His atom completely changes the rules of the game. Nothing, absolutely nothing, in the future research of physics will ever be the same. And it also changes the way we all think about matter. All matter, including the matter we are made of. The idea of solidity vanishes forever. The idea that had guided the first atomists and then scientists of the caliber of Newton. And who could have imagined that the things we touch, that we eat, that we bump into, that we rest our feet on... and our own bodies have no substance, but are the visible expression of waves? To put it differently: we have been forced to change our idea of the "consistency" of matter. It's all, and only, a matter of relationships, of interactions. The floor on which we rest our feet is empty. We are empty. Why don't we penetrate the floor? Be-

cause the floor is an immense structure of electromagnetic interactions, which repel the equally immense structure of electromagnetic interactions that is our body. And if we bump into it, since the interactions that hold the floor together are much stronger than those of our body, it is we who get hurt.
One might say that matter has nothing material about it. But we use words according to old paradigms. We must accept the idea that matter is an expression of energy. And that energy is vibration, wave. And it makes you smile, and it's a smile full of emotion, to think of the monks/philosophers who, by candle-light, reasoning and praying (praying to God to give them the strength to understand the wonders of His world), wrote on their precious sheets that it is light that is the deep reality of things. And that forms are an expression of light. And that the Incarnation is the descent of light upon things. And that after the Incarnation all things are light, as John Scotus Eriugena, the Irish monk/philosopher who lived in the depths of the Middle Ages, between 815 and 877, wrote: "Now that You have become incarnate, even this stone and this wood are light to me".
De Broglie's atom is a child of quantum mechanics. The experimentally observed characteristics could not be explained by the laws of classical mechanics, that of Newton. Quantum mechanics is so important in the development of contemporary physics that it cannot be discussed briefly. Just as one cannot "say a few words" about relativity, the other pillar of contemporary physics. The path proposed in this book is structured to lead to an understanding of the great importance of a very recent discovery: the Higgs boson. We cannot stop.

X. LIGHT

De Broglie's atom is a revolution no less important than the Copernican revolution of the sixteenth century. The idea that particles of matter are waves is even more difficult to believe than the fact that the Earth revolves around the sun. Humanity took centuries to accept Copernicus' heliocentrism. Today we have much more efficient means of information. But we still struggle greatly. It is the weakness of our imagination that hinders us. If all the scientists tell us that it is so, we believe them. But imagining concretely what they tell us is another story. Let's approach the knot in another way. Let's talk about light. And let's see what men have thought about light in the past. It is very interesting and constitutes a different way to understand contemporary physics. It is fascinating what Aristotle thought about light. Far from the truth but fascinating. Aristotle is a philosopher indissolubly linked to the "corporeality" of the universe. He thought that our body contains a certain amount of material fire. The same fire that the sun is made of. He thought that the fire within us is a "hot and compact igneous flow". He thought that every time we open our eyes the hot and compact igneous flow comes out and reaches things. That's why we see them. He also thought that the hot and compact igneous flow that comes out of our pupils is delicate, weak. It couldn't make us see things on its own. So he thought that the light of the sun was needed to allow us to see. The flow from the pupils, igneous and compact but weak, mixing with the powerful luminous flow of the sun, etc. etc. Isn't it fascinating to imagine that light comes out of our eyes? We often say to our women: "bright eyes". If we step out of the straight paths of science for a moment, we can indulge in the poetry of imagining the thousands of rays of light that, in the magical half-circle of stone that was the theater of Dionysus, came out of the eyes of the Athenians and, arriving at the skenè, realized the event of the apparitions. What light was for the philosophers of the Middle Ages we have already seen. I would only add here a quote from a great thinker of that period, Bonaventure of Bagnoregio, a theologian of the thirteenth century. In his Itinerarium mentis in Deo he analyzes at a certain point the process of vision: "The learned image is a similitude, which is generated in the space between the object and the senses, which is then imprinted on the sense organ and, with its imprinting on it, leads us to the knowledge of the principle from which it emanates, that is, of the object. Through this process, the image makes us understand in a manifest way how the eternal Light generates from itself an image or a splendor perfectly equal, consubstantial and coeternal to itself". Bonaventure says that light brings to our senses a "similitude" of the object. The similitude is imprinted on us and makes us know the real object, of which we form, thanks to the image, the "idea". Our knowledge of the real is determined by the images (the "similitudes") that of the real come to us through light. The supremely real is God because he is the light itself that allows knowledge, that "reaches us" through the vision of all things. Bonaventure is fascinating too! In the Middle Ages, light was a fixation. A century before Bonaventure, another theologian, Robert Grosseteste, had written an entire treatise on the subject, entitled De luce. Grosseteste was a professor at Oxford. An English professor, like Newton and Hawking. Even Gothic architecture responded to the needs posed by the "metaphysics of light" and Dante's ascent towards the vision of God is all a vibration of ever purer light until the "Intellectual Light full of love", the goal of every desire. Those whom we contemptuously call "the dark ages" put light at the center of everything. But to talk about science as we moderns understand it, we must start from the seventeenth century. In the seventeenth century, the century of science, there was a debate about the nature of light. Two doctrines were vying for the field. The wave theory and the corpuscular theory. The wave theory was supported by Christiaan Huygens, a Dutchman. The experimental evidence, he argued, tells us with certainty that light is made of waves. The effects of diffraction and interference do not allow us to think otherwise. Newton, the great Englishman, father of classical mechanics, was convinced instead that light was made of corpuscles. He based his certainty, also on incontrovertible experimental evidence. So what? It took us three centuries to understand who was right. And the surprising thing is this: they were both right! Light is made of corpuscles that behave like waves! In fact, even photons, which are the carriers of electromagnetic energy of all kinds, including visible light, behave like de Broglie's electrons. They are particles, but they are also waves. In the twentieth century, a lot of experiments were done on the nature of light. Technology has developed highly reliable and precise instruments. Some experiments say incontrovertibly that light is made of particles. Other experiments say incontrovertibly that light is made of waves. Once again, nature makes a fool of human categories and forces our poor brain to completely change its way of thinking.

XI. MASS AND ENERGY

Photons and gluons, the force-carrying bosons, are privileged particles; they have no mass. They can travel, in free space, at the maximum speed allowed by nature. But they are the only ones. All other particles have mass. However, it is time to clarify what the mass of a particle is. And also of larger bodies. This clarification is necessary because things are not as they seem. Common sense, the result of our daily experiences, tells us that the mass of a complex object is the sum of the masses of its parts. A chair weighs as much as all its atoms put together. It seems indisputable. In fact, the weight of the chair is determined by its mass, that is, by the amount of matter it contains. (However, we must open a brief parenthesis to better clarify one thing: weight and mass are not the same thing. As we have already seen, the weight of a body is the measure of the Earth's attraction on it. It is true that the Earth attracts it in proportion to its mass, but if there were no Earth's attraction nothing would have weight. In space, astronauts float. So the weight of a body is the relationship of attraction that is established between the body itself and a larger body. In other words, weight is not an intrinsic quality of bodies. Mass, on the other hand, is an intrinsic quality of bodies. It measures the quantity of matter of which the body is made. Astronauts in space have no weight but they have their usual mass, which up there does not interact with bodies large enough to attract them. More precisely, we can define mass not as "how much matter" a body has but "how much inertia" a body opposes to changes in motion. This distinction will be useful to us shortly to understand the title of an important article by Einstein in 1905. We will see that things are even more complicated, but for now let's be content with this distinction between weight and mass).

Let's see what happens at the atomic level. A proton contains two up quarks and one down quark. We take it for granted that the mass of the proton is the sum of the masses of the quarks that compose it. There's nothing else in there! But, surprise! If we add up the masses of the three quarks, the total turns out to be only 1% of the mass of the proton. That's right. 99% of the proton's mass is unexplained. The same with neutrons: one up quark and two down quarks. Here too, adding up, etc. etc. If things were as Newton said, that is, that mass is the measure of the quantity of matter (matter understood as he understood it), by adding up the masses of the quarks we should find the mass of the proton. So what? Einstein gave us the answer. In 1905 he was twenty-six years old, like Heisenberg when he wrote down his famous "uncertainty principle". That year Einstein, employed at the Bern patent office, published an article whose title was a question: "Can the inertia of a body depend on its energy content?" The answer is: yes. Einstein showed that energy and mass are interchangeable. Let's take his famous formula:

$E = mc^2$

(energy equals mass times the speed of light squared).

If this formula is true, (and it is true), then it is also true that:

$m = E/c^2$

(mass equals energy divided by the speed of light squared).

These two simple little formulas contain one of the most incredible properties of nature. Let's go back to the proton. The quarks, we have seen, constitute one hundredth of the mass of the proton. The rest of the proton's mass, Einstein's formulas tell us, is the energy that holds the quarks together. Mass does not only measure how much matter is contained in the proton, but is the sum of the matter and the energy that is needed to hold the matter together. It turns out that the (hypothetical) masses of the u and d quarks are rather small, between 1.5 and 3.3 MeV, and between 3.5 and 6.0 MeV, respectively. The mass of the proton is measured at 938 MeV, that of the neutron at 940 MeV. The sum of the masses of two u quarks and one d quark would amount to something like 6.5-12.6 MeV. So where does the rest of the proton's mass come from? It comes from the energy of the gluon fields inside the particle itself. The force that binds the quarks inside the proton (and the neutron) is enormous. The formula $E = mc^2$ tells us precisely that the energy contained in the mass is frighteningly large. When mass is transformed into energy, its value is multiplied by the speed of light squared! Be careful: the principle applies in general and not only to protons and neutrons. Everything that exists, including us, contains a gigantic energy (which makes mass) necessary to hold it together. We thus discover that matter is not that inert thing that the ancients, and Newton, imagined. Matter is solidified energy. When Einstein's article came out, everyone was stunned. All the physicists, I mean. Almost none of them were willing to accept the novelty. But modern particle accelerators have shown that Einstein was right. When two particles are slammed together at very high speeds, what happens is that the set of fragments weighs more than the original particles. The energy released when the particles collide is almost instantly converted into new matter. In short, if I break the chair into pieces and then weigh the fragments, do they weigh less than the whole chair? Of course not. To release the energy contained in the chair, I have to break its atoms and its nuclei and its protons and neutrons. That's where the "strong force" is hidden and acts. But I can't do it. Nobody can do it. Fortunately, because with the mass of a chair all transformed into energy I could destroy the entire city. And even more. The bomb that destroyed Hiroshima unleashed the energy contained in the atoms of just one kilogram of uranium. On the morning of August 6, 1945, the US bomber called Enola Gay dropped a 4-ton bomb on the port city of Hiroshima on the island of Honshu, the largest in Japan. The bomb fell for 43 seconds before exploding. A small conventional charge located near the tail fins fired a projectile of highly enriched uranium into a second block of uranium inside a hollow metal cylinder. Upon impact of the two masses, a beryllium-polonium "initiator" injected neutrons into the uranium. The chain reaction was triggered. The first neutrons produced the fission of a few uranium nuclei. But the neutrons released in the first fission split other nuclei. Other neutrons were released. Each fission released the energy that holds the nucleus together. The number of neutrons released grew monstrously fast. The energy too. The bomb exploded in the sky 580 meters above the city. The atomic age began. To get an idea of the incredible power of the atomic bomb (and therefore of the force that holds the nucleus together), just think that one gram of Uranium 235 undergoing fission produces the same energy produced by the combustion of three tons of coal. And in the end that gram of uranium has lost one thousandth of its mass! One thousandth of a gram is one millionth of a kg. The speed of light is three hundred million meters per second. $E=mc^2$. Three hundred million squared is equal to ninety million billion, which multiplied by one millionth gives the final result of 90 billion. The energy produced is therefore equal to 90 billion joules, or 21.5 million calories. Releasing the energy hidden deep within matter was a staggering achievement. Amazing and frightening. All the physicists in the world are deeply shaken by it. They ask themselves ethical questions. What have we done? Some say. We had to do it, others reply. If we hadn't done it, Hitler's scientists would have.

XII. THE MOST BEAUTIFUL EXPERIMENT

The fact that electrons have a dual nature, being both particles and waves at the same time, is now a proven fact. It is a scientific truth, certified by experiments. The most spectacular experiment in this regard is the one that has gone down in history as the "double-slit experiment". The history of the double slit is long. It begins as early as 1801. As we have seen, Newton thought that light had a corpuscular nature. At the beginning of the nineteenth century, however, the wave hypothesis seemed to triumph definitively. Thomas Young demonstrated that light exhibited phenomena typical of waves, diffraction and interference. Diffraction and interference concern every type of wave, from sea waves to acoustic waves. Now Young's experiment seemed to close the question forever: light too is a wave.

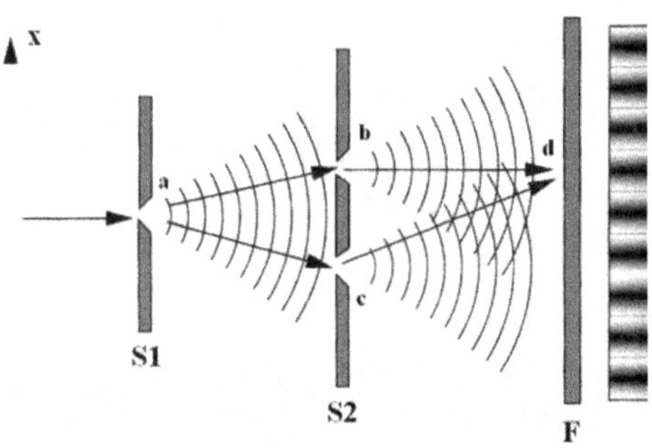

Figure 21

Young's experiment, 1801. Light waves, like waves propagating on water and sound waves in air, interfere with each other when they meet. Due to interference, two crests that meet add up, while a crest and a trough cancel each other out. In the experiment conducted by Young, the light passing through two slits produces alternating light and dark bands on a screen due to the interference of the light waves.

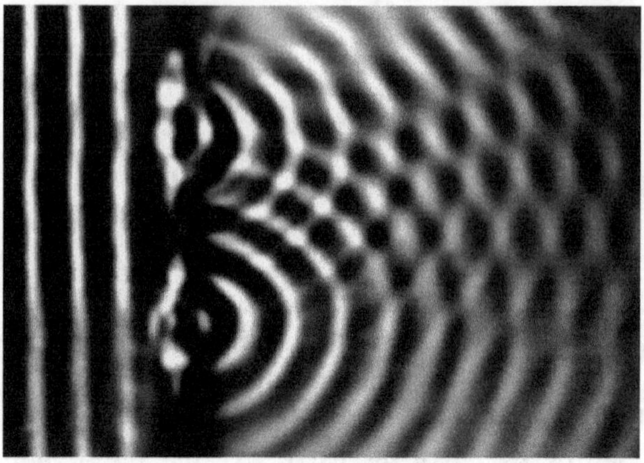

Figure 22

Interference of circular waves obtained with a ripple tank. Brighter areas indicate points where waves add up. Darker areas indicate where they subtract, canceling each other out. Photograph.

Later, James Clerk Maxwell wrote the equations that unequivocally formalized the truth: light is made of waves. But at the beginning of the twentieth century, that truth was once again questioned. In 1905, Einstein provided an explanation of the photoelectric effect.
(Photoelectric effect: a metal plate struck by light—specifically UV rays—produces electricity, i.e., emits electrons; see chapter VII).
Before Einstein, the phenomenon was not well understood. Einstein realized that the extraction of electrons from the metal could only be coherently explained by hypothesizing that electromagnetic radiation consists of energy packets or quanta, the photons. For many years, Einstein's hypothesis was not accepted. Hendrik Lorentz, Max Planck, and Robert Millikan (Nobel Prize winners in physics in 1902, 1918, and 1923, respectively) were convinced that photons did not exist in reality. In interference phenomena, they said, electromagnetic radiation behaves like waves, not particles. And this was unequivocal.

But subsequent experiments proved Einstein right. In electromagnetic radiation (light, heat, etc.), energy is not uniformly distributed across the entire wavefront but is concentrated in discrete packets, energy quanta. A single photon, not the entire wave as a whole, interacts individually with an electron, transferring its energy to it. For this to happen, the single photon must have enough energy to break the electric bond holding the electron to the atom. Thus, in 1921, Einstein won the Nobel Prize, not for the theory of relativity, but for his studies on the photoelectric effect.
And what about the wave?
We face the same problem as with the electron. And the answer is the same: the photon is both a wave and a particle. Not only that. In 1924, De Broglie, as we have already seen, hypothesized that all matter exhibits the same duality.
Proof of the strange behavior of light, and everything else, comes from the experiments of Compton and Davisson and Germer. Compton demonstrated that light behaves like a particle: when a photon is fired at an electron, a perfectly elastic scattering phenomenon is observed. In practice, the photon bounces off like a ball. Davisson and Germer, in turn, fired electrons at low speed against a nickel crystal to measure the angle of incidence of the reflected electron. The electron behaves like an X-ray, that is, like a wave. Thus, quantum mechanics moved from the abstraction of mathematics to the concreteness of laboratories. Planck himself, one of the fathers of quantum mechanics, initially thought that it was just a matter of mathematical subtleties, not reality. As we have seen, he did not share Einstein's hypothesis about the photoelectric effect. But now there are no doubts. Matter behaves this way.
In more recent times, the double-slit experiment, the "most beautiful" experiment, has been conducted again following Young's classical scheme, where an opaque barrier with two parallel slits is placed between a light source and a photographic plate. Now, however, modern detection plates, much more sensitive than those of the 19th century, are used. And the

source is also very different: a very weak source of photons or electrons, or even a source capable of emitting a single photon or electron at a time. At first, it is observed that the plate detects individual bright points typical of particle behavior. However, there is already something very interesting: the points do not correspond to the slits, as would be expected if they were "matter balls," but appear scattered in a chaotic manner. As the number of points increases, interference fringes typical of wave behavior become more evident. This unequivocally demonstrates the existence of wave-particle duality, both for matter and electromagnetic radiation.

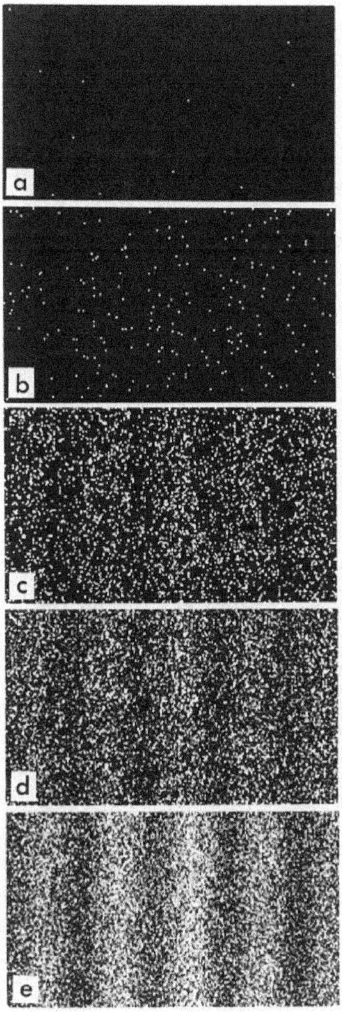

Figure 23

Double-slit experiment conducted with electrons. The images are taken after sending 10 (a), 200 (b), 6000 (c), 40000 (d), and 140000 (e) electrons. The light and dark bands clearly indicate that interference is occurring (see Young's Experiment, Figure 21). However, when the particle is "observed" by being detected on the screen, it appears solely as a particle. Therefore, demonstrating the interference phenomenon is only possible through the observation of many particles.

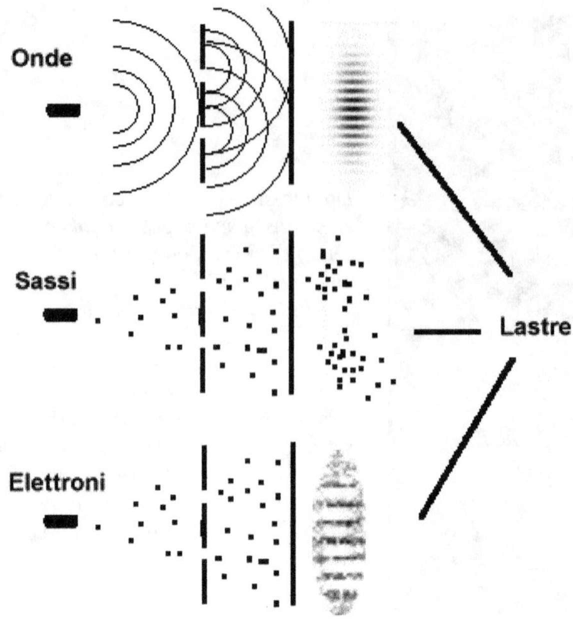

Figure 24

If electrons were simply material particles, they would behave like stones. Instead, they are particle-waves.

It was the German physicist Werner Heisenberg who, in 1927, analyzing Bohr's atomic model, understood its limitations, particularly regarding the orbits of electrons. Talking about orbits implies knowing both the position and the velocity of electrons at specific instants of their motion. Let's imagine we want to determine the position of an electron. We can only rely on photons. To locate an electron, we have to hit it with a photon that is then deflected toward the observer. However, the photon, by interacting with the electron, transfers energy to it, altering its velocity and direction. If, to avoid this problem, we chose to use a low-energy photon, the situation would not improve much. This is because a low-energy photon means a longer wavelength (wavelength and energy are inversely proportional). And the greater the wavelength, the lower the precision. This is exactly what the "uncertainty principle" states.

A consequence of Heisenberg's uncertainty principle is that it no longer makes sense to speak of electrons orbiting the atom along specific paths with a certain energy because it is impossible to precisely define both the orbit and the energy at the same time. A new type of mechanics is needed to provide a more satisfactory description of atomic systems: "quantum mechanics" or "wave mechanics" by E. Schrödinger. In this theory, the ordinary laws of motion are replaced by equations representing probabilities. These probability equations are of the same type as those describing the motion of a point on a vibrating string. They are called wave equations.

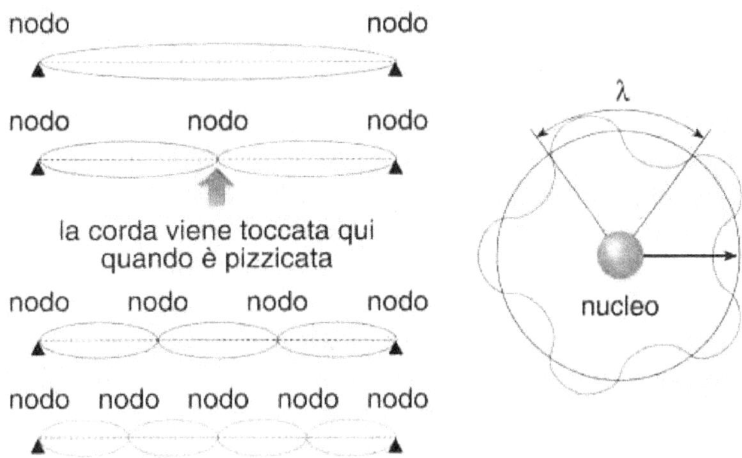

Figure 25

On the left, stationary waves on a vibrating string. On the right, a stationary wave on an orbit with radius r. The complete wave must necessarily contain an integer number of wavelengths λ. In this case, there are five (n=5).

XIII. THE ELECTROMAGNETIC FIELD

We all deal with fields. Not only in the sense that, even though we cannot see them, we are all immersed in fields, but also in the sense that we are aware of one of them. We know it exists. "There's no signal," we say, looking at our cell phone screen when we can't make a call.

On our brief journey toward the Higgs boson, a fundamental stop is the concept of a field, because, as we will see, according to Higgs, the existence of matter is due to the action of fields.

Michael Faraday, a British physicist of the 19th century, conducted experiments that demonstrated the existence of electric fields and magnetic fields capable of interacting. The aforementioned Maxwell derived very accurate equations from these observations. What is it about? An electric charge generates an energy field around itself that exerts a force on any electrically charged object nearby. Experimentally, this can be observed through the attraction or repulsion between bodies with electric charge. A charge is defined as positive when the object has a deficiency of electrons, and negative when there is an excess of electrons. Bodies that are both positively or both negatively charged repel each other, while bodies with opposite charges attract.

But the truly fascinating discovery made by Maxwell is that the electric field and the magnetic field are two sides of the same coin. Under certain conditions, one generates the other, and vice versa. And they act together. A moving magnet creates an electric field. And that's just the beginning. The newly induced electric field, in turn, generates a magnetic field, which then creates another electric field, and so on. These ripples in the electric and magnetic fields, or these variations in their intensity, induce each other and propagate through the surrounding space.

Figure 26

With a "leaf" electroscope, you can observe the electric field in action. It consists of a glass jar containing a metal rod that ends in two very thin gold leaves, the "leaves." When the rod is touched with a charged object, the leaves also become charged and create an electric field between them. Since they have the same charge (that of the object that touched the outer part of the rod), they repel each other. The more charged the object touching the rod, the stronger the field and the greater the distance separating the leaves. These relationships are governed by Coulomb's law, which is very similar to Newton's law of gravitation: the force of the field is directly proportional to the product of the two charges and inversely proportional to the square of the distance.

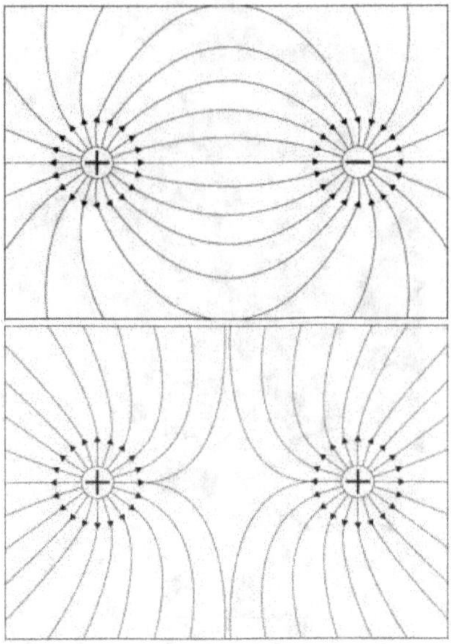

Figure 27

By visualizing the electric field lines, it becomes clear why opposite charges attract and like charges repel. The field lines show the direction and strength of the electric field: they emerge from positive charges and enter negative charges. For opposite charges, the field lines converge, indicating attraction, while for like charges, the field lines spread apart, illustrating repulsion.

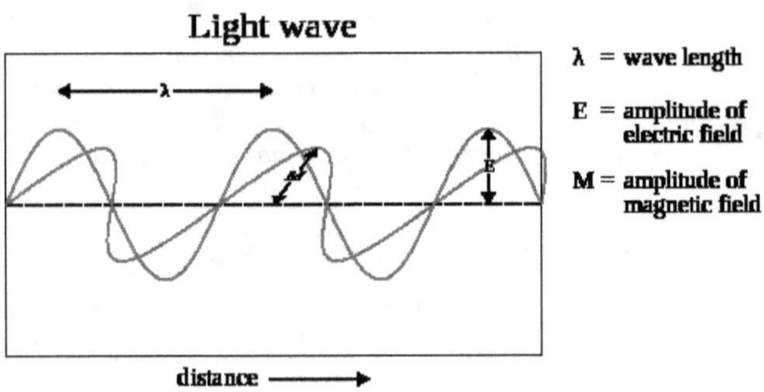

Figure 28

According to the wave theory formalized by Maxwell, the electromagnetic field consists of the electric field and the magnetic field oscillating in perpendicular planes..

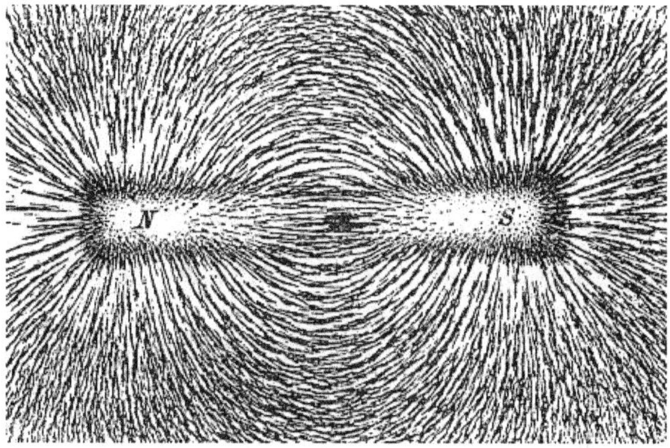

Figure 29

Magnetic field lines made visible by iron filings on a sheet of paper. The magnetic field is generated in space by the motion of an electric charge or by a time-varying electric field. Together with the electric field, it forms the electromagnetic field, which is responsible for electromagnetic interactions. Physical reality is singular: the electromagnetic field, of which the electric field and the magnetic field—described by different Maxwell equations—are merely seemingly separate manifestations. N (north) and S (south) indicate the poles of the magnet.

Maxwell was a curious individual. He wanted to know at what speed the "ripples" of the field propagated outside the source. The answer left him speechless: they traveled at the speed of light. These "ripples" in the electromagnetic field are, in fact, nothing other than light. Einstein would later recognize Maxwell's unique contribution for opening scientists' eyes to the importance of fields: "This change in our conception of reality is the deepest and most fruitful that physics has experienced since the time of Newton."

XIV. THE PHOTON

Fields and force particles are intrinsically connected. Particles are "disturbances" in fields. One could say that particles create fields, as they are their substance. But one could also say that fields create particles—without drops, there is no sea, but without the sea, there are no drops. Photons are the constituent particles of the electromagnetic field. To simplify, we can say that light is made of photons. This is true in the sense that light is composed of electric and magnetic fields, made up of photons, which are the points where the energy and momentum of the field are concentrated and propagate through space as waves. The concept of the photon was introduced in quantum physics to address the contradictions that emerged between classical electromagnetism and experiments conducted at the turn of the 19th and 20th centuries. We have seen that, according to the classical theory developed by Maxwell, light, radio waves, and UV rays were considered electromagnetic radiation, that is, electric and magnetic fields propagating through matter and space following wave dynamics. This is true but doesn't fully capture the phenomenon. Between 1900 and 1905, Max Planck and Albert Einstein introduced the "quantum" of electromagnetic energy as a fundamental indivisible entity, possessing both wave and particle properties: wave-particle duality. This changed everything.

We should never imagine the photon alone, but always in rich company. The Sun emits billions of billions of billions, etc., of photons per second. Experiments have been conducted to count photons. In a dark room, a one-watt light bulb of a single color, green, placed one meter away, sends one trillion photons per second into each of the two pupils of an observer, with a diameter of 2.5 mm. These energy quanta are perceived by our brain as green light. Before Newton, no one had ever thought that light could be made of tiny grains of energy. There are so many of these grains, and they are so close to one another, that they seem like a single, compact entity—a continuous fluid like the one described by Aristotle. So, the photon has no mass, it doesn't weigh anything, and it moves freely, bouncing off the objects in the world. Its limitations are its speed, which, though very high, is not infinite, and its straight-line path. A photon cannot bend. It can bounce off a surface, which is why we see things, but it cannot curve. We see the Moon because the photons emitted by the Sun, which we cannot see directly at night, bounce off the lunar surface and reach the Earth's surface as reflected light. And that's a good thing. If photons could curve, sunlight would reach us even at night, and life cycles would be disrupted. We see sunrise and sunset because the photons emitted by the Sun scatter off atmospheric particles and reach our eyes in a less compact formation than during the day. Some might think it would be fun to see around corners. If light could bend, it would be possible. But then we wouldn't be able to know where objects truly are. Imagine what would happen on the highway. Our brains are used to knowing that objects are where we see them, directly in front of our eyes. If this certainty were lost, we would have no way to move among objects without bumping into them. Even the most advanced instruments can only "assume" that objects are where the light comes from, which allows us to see them.

However, there was a time when people saw things where they weren't. No, it wasn't about drunkards. It was experimental physicists who wanted to finally determine whether Einstein was right. The theory of general relativity was published in 1915. Among the strange predictions it made was the peculiar one about the curvature of space. Einstein said that in the presence of large masses, space bends. The consequence of this curvature of space is that straight lines become curved. They remain straight but curve because space itself is curved. If we walk straight down a curved corridor, we follow the corridor's curvature. Even stranger, something immaterial like light also follows the curvature of space. If a photon could think, it would believe it was moving in a straight line even as it spiraled around a black hole before falling into it.

(To be fair, that brilliant man Newton had already intuited in 1704 that "swarms" of light corpuscles should be sensitive to gravity. But he said this because he thought light was made of material particles. And there he was wrong.)

The theory of relativity was indeed strange. It was a mental revolution. Very few people understood it, and even fewer agreed with it. In 1919, Arthur Eddington and Frank Watson Dyson led an expedition to Africa to see how things stood experimentally. The Sun, they thought, has a very large mass. If what Einstein says is true, the Sun should be able to bend the space around it. The only way to verify this is to observe whether the rays of light passing close to it bend or not. They had to take advantage of an eclipse because otherwise, the rays of the stars, overwhelmed by the Sun's light, would be invisible. The idea of using the eclipse and starlight came from Einstein himself, who as early as 1913, in a brief letter to George Hale, hypothesized the possibility of proving the "bending of electromagnetic radiation while crossing gravitational fields" by examining the light from fixed stars that appeared close to the Sun. He had also calculated how much he thought they should bend: 1.75 arcseconds. On May 29, 1919, there would be a total eclipse in a region of Africa. It was an opportunity not to be missed. The expedition was organized, which was anything but simple, as it was right at the end of the World War. But Eddington, the most prestigious professor of physics at Cambridge, was in love with relativity. He wrote, "Whether the theory ultimately proves correct or not, it deserves attention because it is one of the most beautiful examples of the power inherent in mathematical reasoning." Amid countless difficulties, the expedition set out. They reached their destination. On May 29, the cameras were set up. At the right moment, when an unnatural darkness fell in broad daylight, with shadows all around and stars visible in the sky, they took the photos. When they examined the plates, everyone could see that

stars were photographed that should not have been visible because, according to precise astronomical calculations, their real position was behind the Sun's disk. Einstein was right.
Einstein was a genius endowed not only with superior intelligence but also with good luck—at least in this case. The analysis of the data collected was not the most precise or significant. A strange case of a solid theory confirmed by shaky evidence. We may owe Einstein's global recognition to Eddington's devotion, who, let's say, was a bit hasty in drawing conclusions. The day after his report to the members of the Royal Society and the Royal Astronomical Society, gathered in a joint session, the newspapers reported the news on their front pages. On November 7, 1919, the *London Times* headlined: "Revolution in Science - New Theory of the Universe." A few days later, the *New York Times* published an article with a title that sounded like a sci-fi movie: "Lights All Askew in the Heavens."

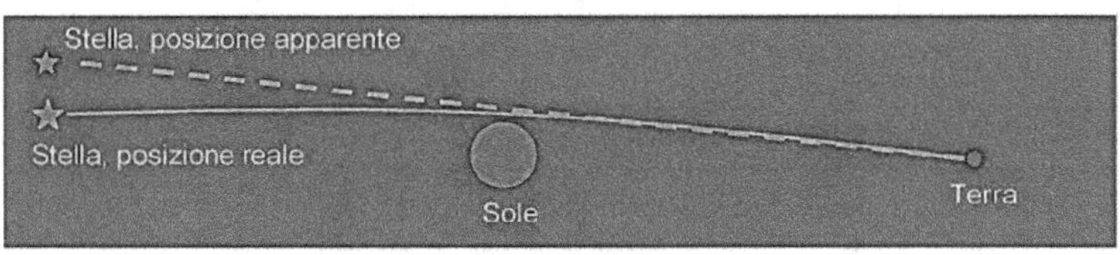

Figure 30

Eddington and Dyson's experiment, 1919. The star in its "real position" should not be visible from Earth. But it is because its light travels straight through space, which is curved by the Sun's mass.

Figure 31

One of the photos taken by the expedition led by Arthur Eddington and Frank Watson Dyson in 1919. The eclipse made it possible to photograph stars that appeared to be close to the Sun but were actually behind it.

Figure 32

Albert Einstein, Vienna 1921. The 1919 Eddington and Dyson experiment instantly made him the most famous physicist in the world. His popularity transcended the boundaries of the scientific community. Everyone knew who he was.

The photon seems to contain the profound mystery of the universe. It has no mass and no electric charge. It appears as an epiphany of nothingness. Other waves, such as sea waves or sound waves, require a physical medium. Sea waves exist because the sea "waves." Sound waves transport mechanical energy through the vibration of air. Without water and air, these waves would not exist. Light waves are different. They do not need a physical medium. They exist in themselves. They are pure energy, pure vibration. In a vacuum, the photon always propagates at the speed of light (we say this, but it is actually the photon's speed that is light). Nothing can travel faster. There is no vantage point from which you could see it moving more slowly.

Let's imagine driving a car at one hundred kilometers per hour and being overtaken by a car traveling at two hundred kilometers per hour. If we aim a speed radar at the passing car, it will measure one hundred kilometers per hour—the speed of the overtaking car minus the speed of the car with the radar. And if that car were coming in the opposite direction, the radar would measure three hundred kilometers per hour, the sum of the two speeds. This is the elementary form of relativity.

Now imagine astronauts on a hypothetical spaceship traveling at 299,000 kilometers per second. If they measured the speed of a light wave moving in the same direction as the spaceship (a laser beam, for example, shot from a station stationary relative to the spaceship), they would find it travels relative to them at 299,792.458 kilometers per second, exactly the same speed as the beam "in itself." And if photons were coming from the opposite direction, towards the spaceship, the measurement would yield the same result. The speed of light cannot be added or subtracted. It is, as they say, "absolute"—free, unbound from any context. Isn't that strange? Yet it is so, proven experimentally many times.

The photon's range is unlimited. It can travel through space-time indefinitely without any limit, until it is absorbed by another particle. At that speed, time does not pass! The photons generated by the Big Bang are still as young as they were then.

Einstein explained to us that if a body is stationary, all its motion is in time. If it moves, part of its motion is in space and part in time. To understand this, it is necessary to have a clear and concrete idea of space-time. There is not space on one side and time on the other. They are the same thing: a four-dimensional reality. We are confused by the fact that we use two words to refer to this reality. Saying "space-time" is the result of the fact that before Einstein, they were thought to be separate things. Just as "electricity" and "magnetism" were before Maxwell. But here our imagination is called to a greater effort, because our experience has convinced us that we know perfectly well what space and time are.

Let's start with a statement that seems absurd: all bodies move through space-time at the same speed, the speed of light. It seems absurd because we cannot imagine movement—and therefore speed—except in three-dimensional space, and we assign time a secondary, controlling role. If we can concretely imagine space and time as one, we can understand how moving at a certain speed means moving in a certain way, in different proportions, through space and time. Space-time is an elastic reality, like a rubber band. If you stretch it lengthwise, it narrows widthwise. The more you stretch it, the narrower it gets. If you stretch it enough to make it immensely long, the other dimension disappears. The same happens if you reverse the directions. Similarly, if a body is stationary, meaning its speed through space is zero, all its movement is in time. If it moves at the speed of light through space, its movement through time is zero.

So, if I drive a car or fly in a plane, does my time pass more slowly than that of someone who is stationary? Yes, but by such a tiny fraction of a billionth of a second that it is in no way measurable, let alone perceptible. Not perceptible, not measurable, but real, because the theory that states it has been proven beyond doubt.

But that's not all. Even the distance from the photon's point of view is zero. This is another incredible wonder predicted by special relativity. The entire distance traveled in the direction of

motion reduces to a single point. The photons from the Big Bang took billions of years to reach our instruments—billions of our years, for us who are stationary and use our speed exclusively to move through time. They have not experienced what we call time at all. They were emitted and absorbed, traversing a universe of zero length in zero time. They shine for an eternal moment, in the absolute.
Moving at the maximum speed allowed by Mother Nature, the photon has zero movement in time. It is because of this property that we can "see" the universe's past, all the way back to its earliest moments, by observing the cosmic microwave background radiation. Those photons were "born" more than thirteen billion years ago and are still traveling ("still" for us; for photons there is no "still" or "before" or "after") without losing any of their original energy. From the time ancient humans looked up at the night sky to now, when we have eyes capable of seeing invisible light—radio telescopes—it has always been these photons telling the story of the world. After the discovery of gravitational waves, we can now "hear" as well as see, but, as the wise Chinese saying goes, "seeing is a thousand times better than hearing."

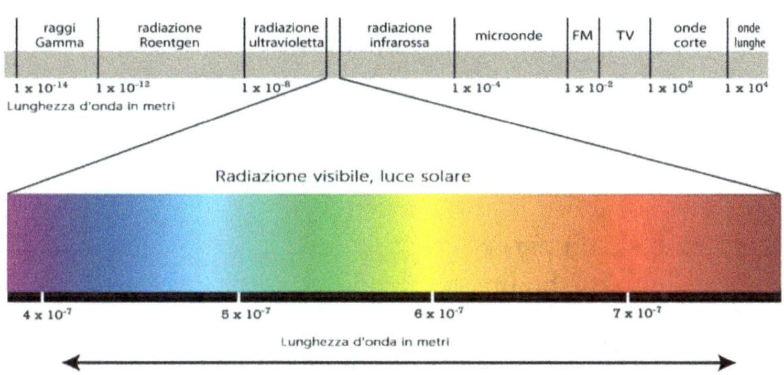

Figure 33

Visible light is just a small part of the electromagnetic field. The rest is light that is invisible to our eyes.

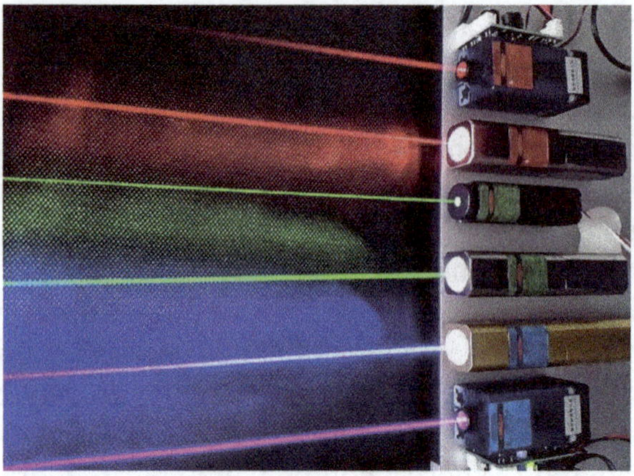

Figure 34

Laser beams are "coherent" bundles of photons. The word "laser" is an acronym for "light amplification by stimulated emission of radiation." In Italian, it translates to "amplificazione di luce mediante emissione stimolata di radiazione." There are many significant practical applications of laser light.

I would like to close this chapter, which talks about stars, the sun, and light, with an excerpt from Dante's *Commedia*, the world-famous Italian poet, who speaks of stars, the sun, and light.

Quando colui che tutto 'l mondo alluma
de l'emisperio nostro sì discende,
che 'l giorno d'ogne parte si consuma,

lo ciel, che sol di lui prima s'accende,
subitamente si rifà parvente
per molte luci, in che una risplende.

Dante says: "When the sun, which illuminates the whole world, slips away from our hemisphere, so that light gradually fades everywhere, the sky, which was previously illuminated only by it, becomes visible again due to the many stars, in which a single light shines." For Dante, stars and planets received their light from the sun. He was mistaken, but what poetry!

Figure 35

Photograph of the Milky Way, the galaxy to which our solar system belongs. ESO (European Southern Observatory).

XV. SYMMETRY

Among all the marvels discovered by twentieth-century physics, symmetry is one of the most fascinating. Symmetry has always been a prominent concept in physics. It was already important in the classical physics of Galileo and Newton and became fundamental in Einstein's special relativity. In contemporary physics, it is a very profound and fertile concept.

We all know what symmetry is. We see it around us in nature, in urban spaces, in our bodies, in houses, in art, in music, and in architecture. It is the basic form of order and understanding. We all appreciate it because it gives us a gratifying sense of harmony and peace. But why is that? Because it is the first form we learn to recognize. The human face is symmetrical. And the human face is the source of all meaning. The survival of human beings is conditioned by their ability to understand the world around them. Understanding the world around them means discovering the logic, the order of things. The first form of order that the human mind discovers is symmetry. If I understand the logic of the things around me, I can protect myself and feel at peace. This is why symmetry is the archetypal form from which all other forms derive. This is why we seek it. Every new space generates some unease and curiosity. The unease disappears when we understand the logic by which the objects in that space are arranged. Understanding means recognizing patterns, as Rudolf Arnheim explains so well in his still-valuable book *Visual Thinking*. Symmetry is the first of these patterns.

Psychology uses symmetry to explore the human mind. It can do so precisely because the symmetrical form immediately stimulates understanding and the attribution of meaning. The famous Rorschach inkblots are based on the power of pure symmetry. They do not mean anything specific. They are ambiguous forms. They do not depict houses, trees, or anything else immediately recognizable. But there is symmetry, and the human mind cannot help but seek meaning. Symmetry demands it.

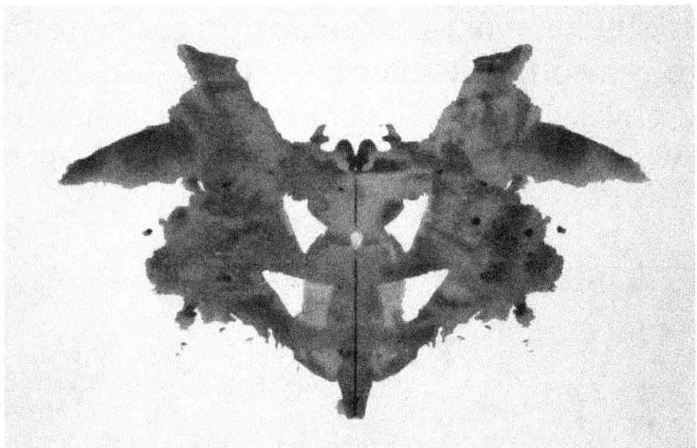

Figure 36

The Rorschach inkblots are shapes that become objects of signification, prompting people to say, "this is..." thanks to symmetry, achieved by duplication along the vertical axis. The patient is asked to instinctively say, without thinking, what the shapes represent. The responses can be indicative of basic perceptual attitudes, revealing aspects of how the subject perceives the world. Rorschach's original work used only form, color, and movement as "determinants." Today, light and shadow are considered another basic determinant. The test using the symmetrical inkblots was created in 1921 by psychologist Hermann Rorschach. The test's validity lies in the ambiguous meaning of the shapes. There is a symmetrical form that stimulates a perceptual response, but what that form represents depends on the personality of the person attributing meaning to it.

In architecture, symmetry is particularly important. Since the Earth's gravity is symmetrical—equal everywhere and at all times—a symmetrical form ensures structural stability. Symmetry has always been perceived as a divine quality. The pyramids were symmetrical, and in the Middle Ages, the era of cathedrals, architecture was humanity's way of imitating the harmony of the world. The "Great Architect," God, created the world according to principles of symmetry. "All things among themselves possess an order," Dante writes. He continues by saying that this order "is the form that makes the universe resemble God." The great cathedrals are symmetrical and are a "resembling form" of the universe, just as the universe is a "resembling form" of God. This idea was so deeply felt as true that architects, to avoid challenging divine perfection, would intentionally incorporate some irregularity—some detail on one side that is absent on the other, a small asymmetry. Because symmetry is perfection, and perfection belongs only to God.

Figure 37

The famous Taj Mahal, one of the most beautiful buildings in the world, was built in 1632 by the Mughal Emperor Shah Jahan in memory of his favorite wife, Arjumand Banu Begum. It is located in Agra, in northern India. It is a perfectly symmetrical structure.

Among all symmetrical shapes, the sphere is the most symmetrical of all. A perfect sphere, without flaws or marks, can be rotated in any direction and will always appear perfectly identical. It doesn't seem to have been moved. Everything looks the same as before the "transformation." This is the concept of symmetry that we all have. A square, for instance, is symmetrical only with respect to 90° rotations, and only if we do not require "point-by-point" symmetry. A rectangle is symmetrical with respect to 180° rotations. As for solids, a cylinder can rotate around its "axis of symmetry" by any number of degrees and will remain symmetrical. However, unlike the sphere, if we rotate it around its horizontal axis, it will retain symmetry only after a 180° rotation.

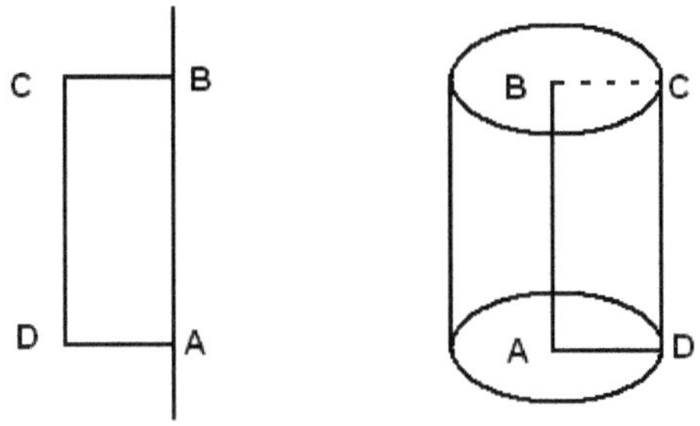

Figure 38

The cylinder is the solid shape generated by the rotation of a rectangle. Its height AB is the axis of symmetry. Symmetry is the expression of "equality between things." These things can be different objects or the same object appearing in the same way before and after some operation we perform on it. In the case of the cylinder, this is true for any rotation around its axis of symmetry. This is an example of "continuous" symmetry".

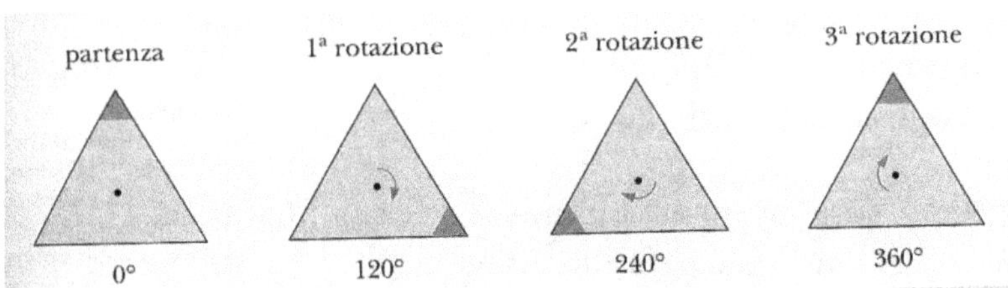

Figure 39

The equilateral triangle has a radial symmetry of order 3. A rotation around its center 0 by an angle of 120° brings the figure back to its original shape, but not point by point. Applying the same rotation 3 times brings the figure back to its original position, point by point. This is an example of "discrete symmetry".

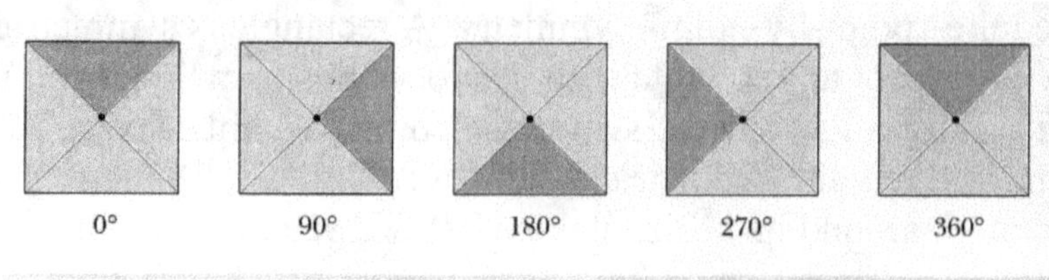

Figure 40

The square is characterized by a radial symmetry of order 4 (four transformations are needed to return point by point to the initial state), centered at the intersection of its axes. Like in the case of the triangle, this is discrete symmetry, meaning it occurs in steps, not continuously.

The examples of symmetry seen so far are easy to understand. They always involve the rotation of an object around itself. However, there are more complex forms of symmetry. The sphere, which we have seen to be the perfect symmetrical object, exhibits complete rotational symmetry. If I move the sphere from one point in space to another, I find that it also has another type of symmetry. Let's keep in mind the definition of symmetry: symmetry is a quality whereby "an object appears in the same identical way after some operation is performed on it." Put differently, and more precisely: "Symmetry is an invariance of a physical system subjected to a change, called a symmetry transformation." So far, we have discussed "rotational symmetry transformations." Now, let's talk about "translation in space." Yes, the sphere has this type of symmetry as well. I can take a billiard ball with me on an airplane or submerge it in a pool. It remains the same. And if I observe the sphere the next day, has anything changed? No. So, the sphere also has this type of invariance (symmetry): it does not change over time. This is called "temporal translation symmetry."
Now let's complicate things and imagine a universe entirely filled, for example, with cylinders. We can imagine rotating them all by the same angle. This would be a "global transfor-

mation." Or we could rotate some of them by one angle and the others by a different angle. In this case, it would be a "local transformation." These concepts will be useful later when we tackle the difficult concept of "local gauge invariance," which is central to the theory underlying the Standard Model.

For geometric figures, which are abstractions, spatial and temporal symmetry are obvious. In concrete reality, things are different. The symmetry of a complex physical system, like a plant or even a single leaf, once identified and described, does not hold over time. In nature, the rule is the continuous breaking of symmetry. We will discuss this further.

Even rotational symmetry, in concrete reality, is like the phoenix. For example, if we rotate a cylindrical vase, we will find its continuous rotational symmetry (Figure 38) only if it has no imperfections. If there is a scratch, the symmetry is broken. But despite this, we can confidently say that nature loves symmetry. It seems to avoid showing it off, as if wanting to hide it in the heart of phenomena, but, if we dig deep, we always find it.

Now, however, we need to talk about a remarkable woman. A mathematical lady. Not only was she a woman, but she was also a "lady of mathematics." Her name was Amalie Emmy Noether.

With her 1918 theorem (actually, it consists of two complementary theorems, one from 1915 and one from 1918, commonly referred to as the "1918 theorem"), Emmy Noether related symmetry to the conservation of energy. Every time we find that a quantity is conserved in a physical process, like electric charge, for example, Noether's theorem tells us that there must be a corresponding continuous symmetry. Noether's theorem thus relates the law of conservation, which is a general law, to the laws of symmetry. The relationship is reciprocal: every symmetry signifies a conservation law. Noether built a bridge between the mathematics of symmetries and physics.

The law of conservation states that every physical system "left alone," meaning undisturbed by external forces, preserves the following in its transformations:

1. Energy.
2. Momentum.
3. Angular momentum.

1. *Energy*: The amount of energy in an "isolated" physical system, that is, "left alone" and not altered by external forces, does not change over time and space, even if there are energy transformations within the system. Respecting this principle led to Pauli's hypothesis of the existence of an unknown particle. Richard Feynman wrote: "There is a fact, or if you wish, a law, that governs all natural phenomena as far as we know them. There are no exceptions to this law; it is exact as far as we know. The law is called 'conservation of energy': we can calculate a certain number, and when we finish observing nature playing its game and recalculate the number, we find that it has not changed."

2. *Momentum*: The momentum of an object is calculated by multiplying the object's mass by its velocity: kg·m/s (kilograms per meter per second), or mass times velocity. A ping-pong ball and a car can have the same speed, but since their masses are very different, their momenta are also very different. In an "isolated" system, the total momentum of the system as a whole, which is the sum of the momentum of its components, does not change over time.

3. *Angular Momentum*: Angular momentum is, simply put, the momentum of a rotating body. The word "momentum" comes from the Latin *momentum*, meaning "movement." Since the radius involved in the circular trajectory, that is, the distance from the center of rotation, comes into play, the formula for momentum (mass × velocity) becomes radius × mass × velocity. This formula explains why an ice skater increases their speed when pulling their arms close to their body: the mass remains unchanged, the radius decreases, and the speed consequently increases. This is called the "law of conservation

of angular momentum."
No experiment has ever revealed the slightest violation of these three fundamental conservation laws. Gravity, for example, is evidence of energy conservation through translation in time. It is always the same. If it were not, we would weigh more on one day and less on another. It also holds for translation in space. Planets would not be round if gravity acted differently on masses in different areas.

Noether's theorem relates the three conservation laws to the three types of symmetry:

1. Continuous symmetry for temporal translation = Energy conservation.
2. Continuous symmetry for spatial translation = Momentum conservation.
3. Continuous symmetry for rotation = Angular momentum conservation.

These are the three great symmetries/invariances of space-time. This is a deep and mysterious relationship that touches on the core issues of physics, such as time and its flow. The relationship between the law of energy conservation and temporal symmetry explains the fact that time intervals do not differ from one another and can flow uniformly in the way we all experience.
That certain invariances must be respected in physics was an obvious fact until the end of the nineteenth century. With the introduction of the theory of special relativity, it became essential that physical laws do not depend on the space-time position of the phenomenon relative to the observer. For laws to be valid, they must be invariant with respect to the observer. If this were not the case, every observer would see phenomena in a way that would be incompatible with what others see, and their own observations would vary from moment to moment and from place to place.
To understand this, it is important to remember that in physics,

“invariance” and “symmetry” are synonyms, as already emphasized: they indicate the same thing. The phenomena described by physical laws must not change when moving through space and time. If they do, those laws are not valid. As Steven Weinberg, Nobel Prize winner in Physics in 1979, wrote: “At the deepest level, all we find are symmetries and their consequences.” This famous statement is quoted in all books about symmetry

In other words, symmetry, in the most complex physics, is a way in which the human mind tries to relate seemingly disconnected phenomena, creating order. The surprising thing is that nature seems to have the same need. Noether's theorem has formalized this reality, putting a powerful tool in the hands of researchers. This means that reasoning through symmetries can lead to discovering new levels of reality.

Emmy Noether was Jewish. She left the University of Göttingen and Germany without ever becoming a full professor. She died in Pennsylvania in 1935. In her obituary in the *New York Times*, Einstein wrote: “Noether was the most significant creative mathematical genius ever produced since women’s higher education began. [...] Pure mathematics is, in its own way, the poetry of logical ideas. One seeks the most general operative ideas that will combine into a simple, logical, and unified form the broadest possible range of formal relationships. In this pursuit of logical beauty, one finds the necessary means for deeper penetration into the laws of nature.” Sheldon Lee Glashow adhered to symmetry, for example, when, dealing with leptons and quarks, he published a paper in 1964 hypothesizing the existence of a fourth quark, which he named charm. This made more sense to him than the asymmetric 4 to 3 ratio, possible with the experimental knowledge of the time. Nature certainly seeks a balance between the number of leptons and quarks: a model with four leptons and four quarks has much more pleasing symmetry.

The most famous proof of the power of Noether's theorem was the theoretical prediction and subsequent discovery of the neutrino. A beautiful story worth telling briefly. The protagonist of this story is the Austrian physicist Wolfgang Ernst Pauli. I will also mention some aspects of his troubled life, just to give the impression that physicists are human beings like everyone else, with private lives, loves, disappointments, and everything else. It would be nice to do this with all the brilliant physicists mentioned, but the size of this book would become excessive. Let's talk about Pauli and take him as an emblem, even though his life was particularly restless.

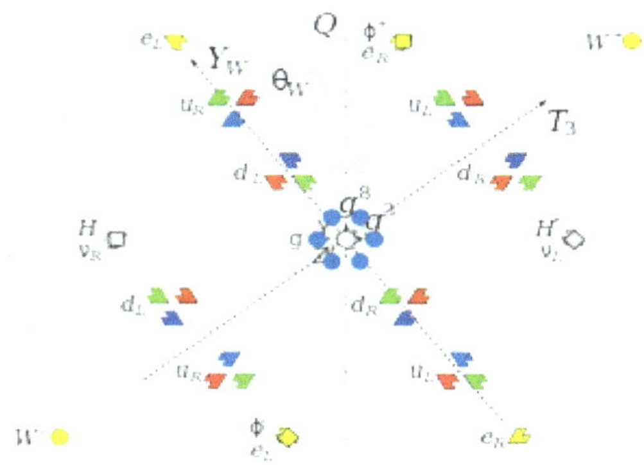

Figure 41

A symmetrical representation of known particles arranged according to their characteristics. When discussing symmetry at the particle level, terms like "symmetry groups," "multiplet," "octet," and "decuplet" are used. Gell-Mann and Ne'eman, for example, noticed that spin-1/2 baryons can be arranged into an octet, as can mesons (see the following figure). Reasoning through symmetries can lead to the discovery of new levels of reality.

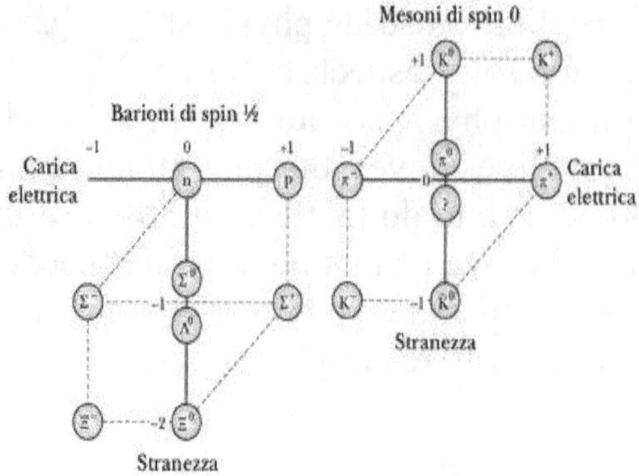

Figure 42

A famous example of verifying the general law of symmetry is known in history by the Buddhist-inspired name "Eightfold Way" (the eight steps to reach Nirvana). In 1961, Murray Gell-Mann discovered that he could arrange baryons, which include the neutron (n) and proton (p), and mesons into two octets corresponding to eight-dimensional representations of the global symmetry group. However, in the representation of mesons, there were only seven particles. There was, in fact, a box with a question mark. The missing particle was discovered a few months later by the group led by Luis Álvarez at Berkeley and was named eta (η). Among physicists, this is referred to as a "return to Plato and Pythagoras".

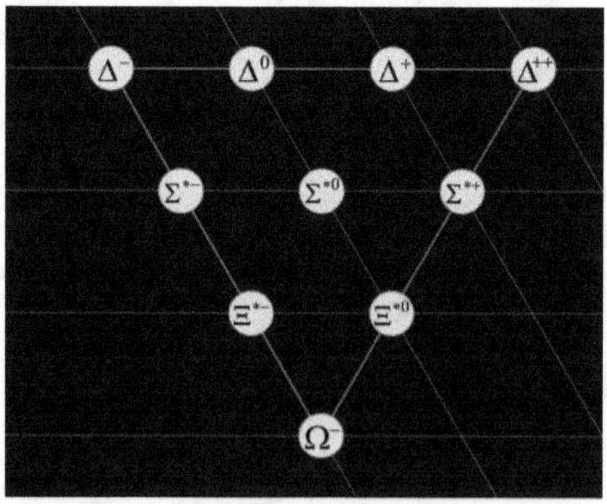

Figure 43

The baryonic decuplet describes the combination of the three quarks: u, d, and s with a total spin of 3/2. It resembles Pythagoras' tetractys. The baryonic decuplet was drawn by Murray Gell-Mann in 1961, at a time when the particle at the top, which was discovered in 1964, was still unknown.

XVI. LIFE AND WORKS OF A TORMENTED GENIUS

Wolfgang Ernst Pauli was born in 1900 in the Döbling district of Vienna. His mother was Berta Camilla Schütz, and his father was Wolfgang Joseph Pauli, an important chemist and university professor. In 1898, his father, who was of Jewish origin, changed his surname from Pascheles to Pauli, converted to Catholicism, and got married in a church. Pauli's middle name, Ernst, was given in honor of his godfather, Ernst Mach, who was also a scientist.

Young Wolfgang Ernst studied at the Döblinger Gymnasium in Vienna, where he graduated in 1918. Just two months after graduating, he published his first paper on Albert Einstein's General Theory of Relativity. In 1921, he enrolled at Ludwig Maximilian University in Munich, where he earned his PhD in physics under the guidance of Arnold Sommerfeld. Fate, or coincidence, had it that his desk mate was Werner Heisenberg, another future genius of 20th-century physics. Professor Sommerfeld asked Pauli to write an article on relativity for the *Encyklopaedie der Mathematischen Wissenschaften*, an important and extensive encyclopedic work. Two months after receiving his PhD, Pauli completed the article, a 237-page manuscript. The size and quality of the work turned it into a monographic book, which was published. Einstein read it and was deeply impressed.

Pauli was an exceptional student, and he knew it. His peers and professors admired him, but he was dismissive of those who couldn't keep up with his pace of thinking. His famous remark about a paper he didn't like was: "It's not even wrong." This attitude stayed with him throughout his life.

(A famous episode illustrates his character. On February 23,

1954, the well-known physicist Chen Ning Yang, originally from China, who had also worked with Fermi, gave a seminar at Princeton where he presented work done with Robert Mills to understand what was wrong with the complex and troubled theory called Quantum Electrodynamics (QED). He had some solutions, but they were not entirely satisfactory. Among the attendees were Oppenheimer and Pauli, who had moved to Princeton University in 1940. It seems Pauli had tried to follow the same logical path and had reached similarly unconvincing conclusions. The issue was with the mass of the force mediators, which, according to those equations, resulted in zero—an absurd outcome. Heisenberg and the Japanese physicist Hideki Yukawa had shown in 1935 that particles mediating short-range interactions, such as the strong nuclear force, must be "heavy." Pauli had therefore abandoned the idea. While Yang was writing his equations on the blackboard, Pauli, knowing where the weak point lay, asked, "What is the mass of this B quantum?" "I don't know," Yang replied softly. "What is the mass of this B field?" Pauli insisted forcefully. "We've thought about it," Yang replied. "It's a very complex issue, and we can't give an answer now." "That's a pretty weak excuse," Pauli hissed.)

After completing his studies, he spent a year as an assistant to Max Born at the University of Göttingen. The following year, he went to the Niels Bohr Institute of Theoretical Physics in Copenhagen at Bohr's invitation. Even Bohr, one of the fathers of atomic theory, was impressed by his abilities.

Accustomed to understanding everything, Pauli encountered a problem stronger than himself during these years: the Zeeman effect. This phenomenon involves the splitting of spectral lines due to an external magnetic field. Each line splits into multiple closely spaced lines, but at that time, it was not understood why. His inability to find a solution led Pauli into a form of depression that would accompany him throughout his life. However, from 1924 to 1928, he was a professor at the University of Hamburg, and these were decisive years during which he contributed to the development of quantum mechanics. Most notably, these were the years of the formulation of the "Pauli Exclusion Principle."

(In physics, the term "principle" refers to an assumption or foundation of a theory that is evidently true. A principle cannot be contradicted; a theory that contradicts it is incorrect.)

In 1925, Pauli formalized the principle that bears his name. In an atom, an orbital can host only two electrons, and they must have different characteristics. Put differently: in an atom, there cannot be two electrons with the same set of four quantum numbers. What are quantum numbers? There are four quantum numbers that describe the characteristics of the electron cloud (atomic orbital, meaning the region within the atom where the electron moves, rotating and oscillating):

1. The Principal Quantum Number "n" indicates the energy and size of each atomic orbital. It can only take integer values: 1, 2, 3... We already discussed this in Chapter VIII, dedicated to Bohr.

2. The Secondary Quantum Number "l" (also known as the azimuthal or angular quantum number) indicates the shape of an orbital. It can take values of 1, 2, 3.

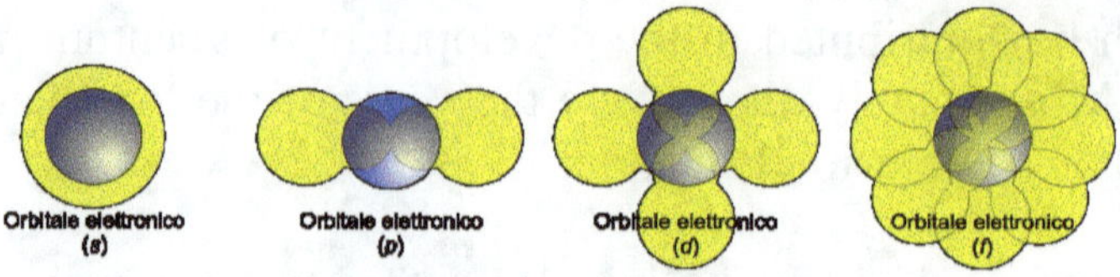

Figure 44

Secondary Quantum Number: For $l = 0$, the electron cloud is spherical (s orbital). The nucleus is at the center of the sphere. For $l = 1$, the electron cloud has two lobes (p orbital), with the nucleus at the center of the two lobes. For $l = 2$, the electron cloud has four lobes (d orbital), with the nucleus at the center of the four lobes. For $l = 3$, the electron cloud has eight lobes (f orbital), with the nucleus still at the center, of course..

3. The magnetic quantum number "m" indicates the number of possible orientations of a given orbital in space and depends on the value of the secondary quantum number l. For $l = 0$ (spherical orbital), m can only be 0. This is because the orbital, being spherical, has only one possible orientation in space. If $l = 1$ (double-lobed p orbital), there are three possible orientations. And so on. See *Figures 45, 46, and 47*.

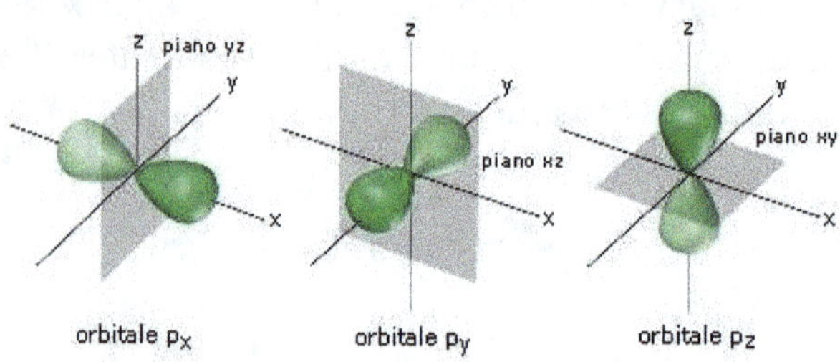

Figure 45

p Orbitals: For $l = 1$ (double-lobed orbital), m takes three values: -1, 0, +1. These three values correspond to three double-lobed orbitals oriented along the x, y, and z axes.

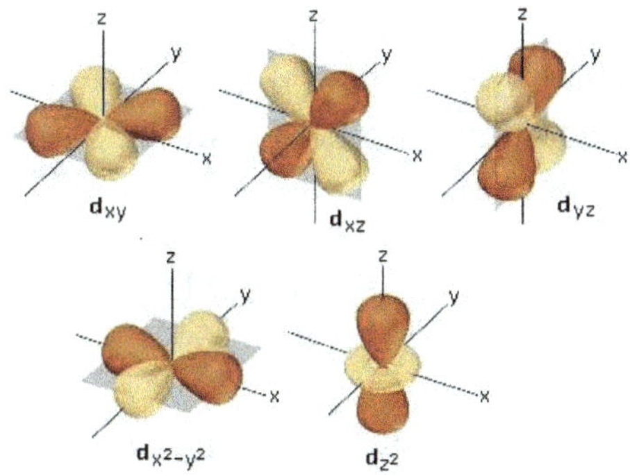

Figure 46

d Orbital: For \(l = 2 \), with four lobes, the magnetic quantum number \(m \) has five possible values (-2, -1, 0, +1, +2), which correspond to five different spatial orientations of the orbital.

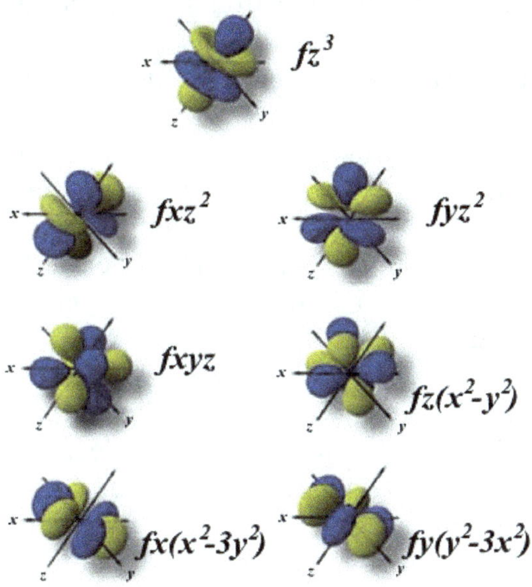

Figure 47

f Orbital: For \(l = 3 \), with eight lobes, the magnetic quantum number \(m \) has seven possible values (-3, -2, -1, 0, +1, +2, +3), representing seven different spatial orientations of the orbital.

4. The Spin Quantum Number "ms"

The first three quantum numbers distinguish between different atomic orbitals, or electron clouds. The fourth quantum number pertains specifically to the electron and can only have two possible values: $ms = +\frac{1}{2}$ and $ms = -\frac{1}{2}$.

In November 1925, two Dutch physicists, G.E. Uhlenbeck and S.A. Goudsmit, proposed that the electron rotates around its own axis, much like the Earth. If this is true—and it is—the electron (like any other material system) produces an effect in the surrounding space when spinning, represented by a vector whose orientation coincides with the rotation axis of the particle. This property is called the "electron's spin angular momentum" or simply the "electron's spin." Since the electron also possesses an electric charge, its rotational motion gives it the properties of a tiny magnet with a north and south pole.

The mathematical formulation by Goudsmit and Uhlenbeck led to the conclusion that the electron can only spin in two directions: one right-handed and the other left-handed, referred to as "spin up" (↑) and "spin down" (↓). However, in 1927, when the English physicist Paul Dirac combined quantum mechanics with Albert Einstein's theory of special relativity, it soon became clear that the visual metaphor of terrestrial rotation had no basis in reality. Today, we interpret electron spin as a purely "relativistic" quantum effect, where the electron can have one of two possible "orientations" that we call spin-up and spin-down. These are not orientations along particular directions in conventional three-dimensional space, but orientations in a special "spin space" that has only two dimensions: up and down.

It is impossible to form a visual image of the electron's two different orientations. Any attempt to visualize them would be incorrect. Although their effects are real and can be studied in detail in the laboratory, there is nothing comparable in the visible world. We must be content to know that spin determines the magnitude of the electron's angular momentum—the physical quantity associated with its rotational motion—and governs how it interacts with an external magnetic field.

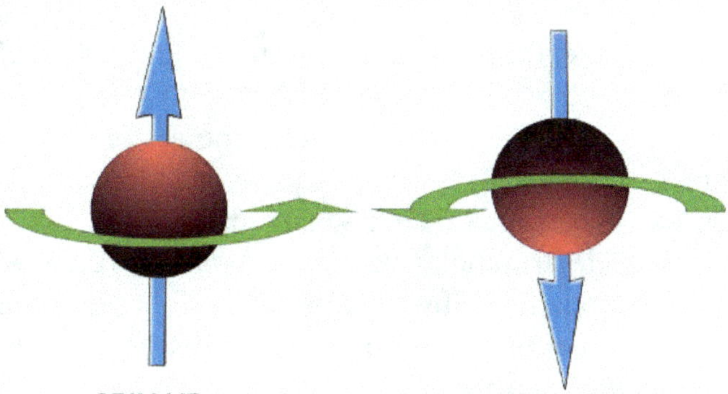

Figure 48

According to the Pauli Exclusion Principle, in a system of particles (such as two electrons in an atom), it is forbidden for them to be in the same energy state, meaning they cannot have the same physical characteristics. This is the only way the structure of atoms can be fully described by the theory. The Pauli principle applies only to particles with half-integer spin (1/2, 3/2, 5/2, etc.), just like the electron, which has a spin of 1/2. In fact, according to the Standard Model of matter, all elementary particles that make up visible matter have a spin of 1/2 and are generally called "fermions", following Enrico Fermi's classification.
Mathematically, this means that their wave function, the equation describing the current state of the particle, is antisymmetric with respect to rotations. That is, for a complete rotation (360°), the wave function changes sign (it is antisymmetric!). It takes another complete rotation (a total of 720°) for the wave function to return to exactly the same state it had before the rotation.
On the other hand, the mediators of the fundamental forces, "bosons", have a symmetric wave function. Consequently, after a 360° rotation, they return to their initial state..

(What is a vector? In physics, there are two types of quantities: "scalar quantities," which can be fully defined by their measurement relative to a predefined unit, and "vector quantities," for which direction and magnitude must be associated with the measurement. Mass, temperature, and time are scalar quantities. They are expressed with a number and a unit of measure. Forces, velocity, and acceleration are vector quantities. In addition to magnitude, they require geometric indications: direction and orientation.)

In summary, the four quantum numbers uniquely define the state of an electron. It is impossible for two electrons in the same orbital to be described by the same set of four quantum numbers, as identical sets indicate the same electron.

As mentioned earlier, in 1925, Pauli established that a maximum of two electrons can occupy the same orbital, and they must have opposite spins. This principle is of fundamental importance as it explains the stability of matter, even on a large scale. Molecules cannot be pushed into each other because the electrons of each molecule cannot occupy the same state as the electrons of another molecule.

In 1928, Pauli became a professor of Theoretical Physics at the Federal Institute of Technology in Zurich, Switzerland. In May 1929, he left the Roman Catholic Church and, in December, married Käthe Margarethe Deppner, from whom he divorced in 1930, after less than a year. From 1930 to 1937, he participated in major physics conferences in Russia, the United States, and Italy. In 1934, he married Franciska Bertram, with whom he remained until his death. The annexation of Austria by Germany in 1938 made Pauli a German citizen—a Jewish German. At the outbreak of World War II, he emigrated to the United States, where he became a professor of Theoretical Physics at Princeton, the leading physics department in the world at the time. After the war, he returned to Zurich, where he lived and worked until his death in 1958.

There is a very particular aspect of Pauli's life. As we have seen, he was a person of great and restless intelligence, prone to depression. He suffered greatly from it. He was unhappy, and despite his scientific success, he often vented his frustration by getting drunk and fighting in Zurich's nightclubs. His mother had committed suicide, and his father had remarried ("too quickly," Pauli thought, like Hamlet). His father's new wife was the same age as Pauli, who saw her as an "evil stepmother." So, Pauli married a cabaret singer but divorced after a few months. "Relationships between women and me," he wrote to his friend Wentzel, "don't work at all. I'm afraid that as I grow older, I will feel increasingly lonely." Concerned about his son's well-being, Pauli's father, a colleague of Jung at the University of Zurich's Medical School, asked Jung to treat his son.

It was a "rich and strange" encounter, as Ariel sings in *The Tempest*. To Jung, who listened to Pauli's dreams, it seemed as though he was witnessing the living confirmation of his theory of the collective unconscious. "Pauli's dreams," he wrote, "are overflowing with archaic material." Jung, not wanting to influence this extraordinary patient with his enthusiasm, referred him to a colleague, suggesting that, given Pauli's issues with women, this would be more appropriate. But he asked Pauli to send him detailed written accounts of his dreams. He received over 1,300 meticulous dream reports from his colleague, material that would prove invaluable for Jung's studies on psychology and alchemy. An intense correspondence, full of intriguing insights, developed between Pauli and Jung. Jung wrote, "It seems illuminating to me that in quantum physics, the archetype must be sought in the concept of probability (mathematical), that is, in the actual agreement between the expected result, calculated with the help of this concept, and the empirically measured frequencies," just as, in psychoanalysis, "the archetype represents nothing other than the probability of a psychic occurrence. To a certain extent, it is the anticipated result, prefigured in images, of a psychic statistic." In 1952, the physicist and the psychoanalyst published the book *Nature and the Psyche*, containing texts by both. In one of these texts, we read: "Experience has shown that both light and matter behave, on the one hand, like separate particles, and on the other hand, like waves. This paradoxical result has made it necessary to abandon—at the atomic scale—a causal description of nature in the usual space-time continuum, replacing it with invisible probability fields in multidimensional spaces, which actually represent the stage reached by our current knowledge on the subject." They sound like the words of a physicist, but they were written by Jung. On the other hand, in the same book, we read: "The division and reduction of symmetry—that's the heart of the beast! Moreover, division is an ancient attribute of the Devil. If the two divine rivals, Christ and the Devil, only knew that they have become so symmetrical!" And these words were written by Pauli.

The multifaceted richness of Pauli's mind is evident in all his writings, including his letters. In 1947, while working on his wonderful book on Kepler, he wrote to his friend Markus Fierz: "I continued my journey through the 17th century. The fact that Newton, so to speak, placed space-time at God's right hand, that is, in the vacant place of the Son whom he had driven away from there, is a particularly tantalizing aspect of the history of Spirit. It took an extraordinary effort to bring space and time down from Olympus again. And this task was artificially made even more difficult by Kant's philosophy, which tried to bar human reason from access to that Olympus. Therefore, I am particularly interested in the period when space and time were not yet up there, namely the phase immediately preceding this fatal operation. Hence my study of Kepler."

Thousands of books have been written about the relationship between Pauli and Jung, and consequently about the relationship between quantum physics and depth psychology. Most of them are full of nonsense. There is no scientific basis for the relationship between physics and psychoanalysis. There is no way to transcribe psychic processes into mathematical formulas. But Pauli and Jung's interest in both fields reflects a search for meaning, which modern philosophy will take into account. Pauli never renounced the principle of objectivity in his science. But this did not prevent him from seeking something "beyond" physics, capable of filling the void left by the disappearance of the *Anima Mundi*. This is a need that many contemporary physicists feel deeply.

Figure 49

Wolfgang Ernst Pauli. Einstein praised him for "the psychological understanding of the evolution of ideas, the accuracy of mathematical deductions, the profound intuition, the ability to present work with systematic clarity, factual completeness, and infallible critical sense." Pauli was deeply engaged with Jungian psychoanalysis and, as director of the scientific committee of the Carl G. Jung Institute in Zurich, advocated for psychoanalysts to undergo analytical therapy themselves. This need, which seemed excessive at the time, is now a well-established practice..

And what about Pauli's neutrino? In the next chapter. This one is already long enough.

XVII. THE DISCOVERY OF THE NEUTRINO

Radioactivity is one of the most studied phenomena in modern physics. Everything that exists emits radiation. Even our own bodies emit infrared rays, or heat radiation, to be precise. Among the various forms of radioactivity is beta radiation, \(\beta \), which is typical of radioactive materials and consists of high-speed electrons. The process that generates \(\beta \) radiation is peculiar. It seems strange, but nuclei with an abundant number of neutrons become unstable and transform to reach a new equilibrium. A neutron in such a nucleus "decays," transforming into a proton and emitting a high-speed electron, known as a \(\beta \)-particle. If, instead, a proton is in excess in the unstable nucleus, it transforms into a neutron plus a \(\beta \)-electron. This second case is known as positive \(\beta \)-decay, while the first case is called negative \(\beta \)-decay. It was later understood that the process is governed by the dynamics of quarks, the building blocks of protons and neutrons. A neutron consists of two down quarks and one up quark. One of the two down quarks transforms into an up quark. However, since the down quark has a charge of -1/3 and the up quark has a charge of 2/3, there is now too much energy in the nucleus. At this point, a "virtual particle" called a \(W^- \) boson intervenes to carry away a charge of -1. The unstable neutron has now changed its nature: it has become a proton. The \(W^- \) boson, having completed its task, moves away. The atom to which the "neutron that became a proton" belonged also transforms and becomes… another element. A spontaneous form of natural alchemy! Meanwhile, the virtual \(W^- \) boson gives rise to an electron and… Today, we know it also gives rise to another particle, specifically an antineutrino, but in 1930, this was unknown, and \(\beta \)-decay was an apparently unsolvable enigma. Experimental evidence clashed with theoretical calculations. Something was missing: a particle with zero mass. This meant the decay process involved three bodies instead of two. However, the technology of the time was not advanced enough to detect this third particle. In physics, if something cannot be "seen," it doesn't exist.

Yet, the alternative to the existence of the unknown particle was to abandon two fundamental principles: the conservation of energy and the conservation of momentum, which adhere to the more general principle of symmetry. This issue caused significant turmoil in the scientific community. Bohr suggested that there might indeed be a violation of energy conservation. Pauli, however, held firm to the principles, insisting that the decay process was not a two-body process, as observed, but a three-body process, meaning that among the products of the reaction, there was indeed a third particle—very small, electrically neutral, and undetectable with the available instruments. It was Pauli who, courageously, chose to uphold symmetry, despite the experimental evidence.

Years passed, but experimental physicists still could not detect the elusive missing particle. It began to be thought that it might not exist after all, even though it had already been humorously named "neutrino," the unknown child of the neutron, by Amaldi and Fermi, as we have seen. But in the end, nature responded positively to Pauli's act of faith in her (and her deep symmetry), and some twenty years later, in 1956, physicists Clyde Cowan and Fred Reines "saw" the neutrino during an experiment conducted at the Savannah River fission reactor.

XVIII. INTERLUDE

I am writing this book in a valley of the Ligurian Apennines. Here, I have a small house, once the home of cattle farmers. I have been completely alone for days. I listen to the wind. I observe the sky's unfolding events. The clouds that rise from the valley and reach the summit here. Some mornings everything is wrapped in white cotton. Other times, it's the blue that dominates—an ethereal, windy blue. What is the fundamental level of reality? I wonder. The electromagnetic field between 390 and 700 nm, or the colors that reach me with such emotion? In front of the house is a large linden tree. It's turning yellow. It's autumn. In the evening, the just-set sun lights up the horizon over there, behind the Alps. At night, the moon and stars are visible like they no longer are in the city. The rest of the world seems far away. How strange we are. It only takes a few days and everything recedes. I am captivated by the poetry of nature, of the passing time, of the color of the leaves, of the wind that stirs the linden tree outside. I know that behind all of this are fields, interactions, particles... nature's magic tricks. As the poet says: "What would life be if, laboriously climbing the hill, one did not see, by the side of the ditch, the daisy saying: look at me, I'm blooming." I also know that everything is governed by an inscrutable symmetry. Inscrutable, if I ask why. And even though I know the mechanisms of conservation and change, I am increasingly aware, as the years pass, that there are questions, and they are the most important ones, to which we have no answers. But it doesn't sadden me. I used to be frightened by the thought that matter doesn't exist and that everything we see is just an appearance, an illusion of the senses. But now I sense the beauty in all of this. The trees, the entire forests around here, the stones, the water, the sky's air, the house's walls are energy. If we think, "they are only energy," nihilism seizes the soul. But all these things, all things "are"—they are energy that for a while condenses into concrete forms, as if wanting to be seen, as if energy decided to pause for a moment. For us. To create our world. To create us. Then it will return to being energy, that is, pure force, the spirit of the world.

The book is taking shape. I write spontaneously, without thinking too much. Sometimes, during the day, I stop, step out onto the terrace, and read Basho, the Japanese poet:

noble is he
who does not deduce from lightning
that nothing lasts

I alternate Basho's haikus with a book by Severino. The Plato of our time. I admire Emanuele Severino's argumentative power, his rigorous reasoning. And above all, I admire his courage. In an era when everything seems destined to disappear quickly, he writes about "eternals." All life, the life of each individual, is a "coming forth of the eternals." Severino starts from the theory of relativity, which asserts that the past and the future have the same consistency as the present. He deduces that change is an illusion. That time does not flow in reality. That we are a circle of light into which the eternals enter, one by one, according to us. And this is time. That everything is like a huge film strip. We see individual frames, one after another, and we have the illusion that they "happen" when we see them. Instead, the immense film exists in itself, independent of our gaze. Things don't happen; they *are* and are forever. Even physical phenomena, the innermost ones, "are." I find it singular, moving, to think, while reading Severino, that in the mathematical formulas studied in certain fields of current physical research, such as quantum gravity, the time variable has disappeared. And I think especially of what relativity tells us: that all objects, including us, move at the speed of light, only that almost all objects, including us, move through three of the space-time dimensions at ridiculously low speeds, which is why our movement is entirely within the fourth dimension, time, whereas for those traveling through the three dimensions at maximum speed, photons, the fourth dimension, time, is completely absent, quantity zero. Is there a relationship, I wonder, between this and Severino's thought, which is based entirely on the negation of time and transformation, including death? Severino's arguments are rigorous, as I said, and very complex. He wrote about fifty books. I enjoy reading one up here; I enjoy breathing the rarefied air of his metaphysics.

XIX. THE BREAKING OF SYMMETRY

In 1961, Yoichiro Nambu (a Japanese physicist who became a naturalized American and would win the Nobel Prize in 2008) from the University of Chicago published a paper attributing the transformation of massless particles into massive ones to a kind of spontaneous symmetry breaking involving the entire universe. His article didn't prove anything, but it planted the idea in the minds of physicists, including Higgs, that symmetry breaking might indeed be the key to the origin of mass. Nambu's theory would later be proven incorrect, but Higgs and others sensed something profound in his research. Without being aware of each other, various research groups in the US and Europe began exploring Nambu's theory, convinced it concealed something very interesting.

But what does "symmetry breaking" mean?

Let's take an example. Consider our perfectly symmetrical sphere. Let's go over this again: it is perfectly symmetrical because no matter how we rotate it, it appears the same. Now imagine it made of metal. Apply some external energy to the sphere—let's say we crush it. At some point, its shape changes. If we keep pressing, we turn it into a disc. Its initial symmetry is broken. It probably still has rotational symmetry around its vertical axis, but only around that axis. The Earth, for instance, rotates around its vertical axis, and if we ignore the surface irregularities, we can say it is symmetric in that rotation. If, however, it were to rotate around an axis perpendicular to its usual one, it would no longer be symmetrical in rotation, because the Earth is flattened at the poles. If that earlier sphere were made of ice instead of metal, we wouldn't need to crush it; just exposing it to a temperature above 0°C and waiting a bit would suffice.

Let's take another example because the concept is not the easiest to grasp. When a top spins at high speed, there is perfect symmetry in all directions: the top rotates around its "axis of symmetry," and each direction is equivalent to any other. As the speed decreases, the symmetry falters. The top wobbles, staggering between different directions. When the top finally stops and rests on the ground, it has chosen a single direction. The initial symmetry is permanently broken. There has been a "spontaneous symmetry breaking." It may seem that the chosen direction is entirely random. In reality, it's due to tiny fluctuations in the surrounding environment, which are the "background noise."

Even the photon, the luminous prince, can be involved in symmetry breaking. In a superconductor at extremely low temperatures, electrons reorganize in such a way that they break electromagnetic symmetry. The paradoxical consequence is that when a photon, which we know is a massless particle of light, comes into contact with a superconducting material, it immediately gains mass. Inside a superconductor, even photons become massive due to symmetry breaking. This phenomenon is known as the "Meissner effect."

Cosmologists have applied this same reasoning to the evolution of the universe. At the moment of the Big Bang, there existed a single force, which we can call the "superforce." All particles were identical, massless, and moved at the speed of light. The universe was, therefore, perfectly symmetrical in every sense of the word. Immediately after, in mere billionths of a second, symmetry broke down due to the drop in temperature to just a few million billion degrees. New forces and new particles emerged. But now we need to take a step back and understand why the recent (2012) discovery of the Higgs boson is so important.

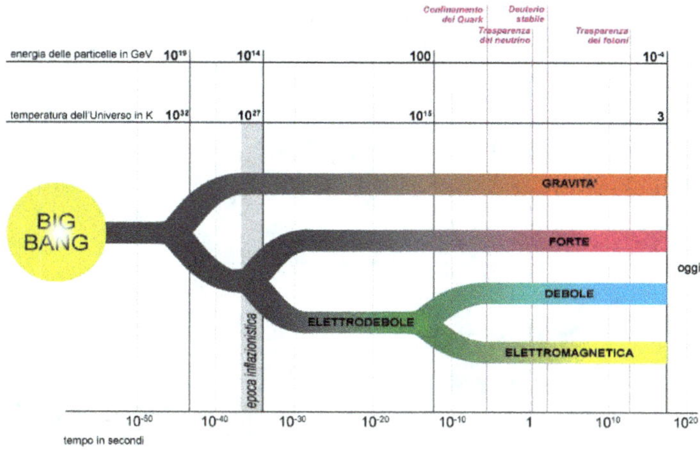

Figure 50

The spontaneous symmetry breaking in the history of the universe, according to the electroweak theory. After one second from the initial explosion, the four forces of nature were active, following two successive symmetry breakings. Recall that GeV means giga-electronvolt and K degrees Kelvin. Today, the temperature of the universe, after thirteen billion and seven hundred million years since the Big Bang, is 3K, or 270 degrees below zero. (Graphic elaboration by F. Cuicchio).

XX. THE MASS OF PARTICLES.

Let's take the Standard Model in hand again.

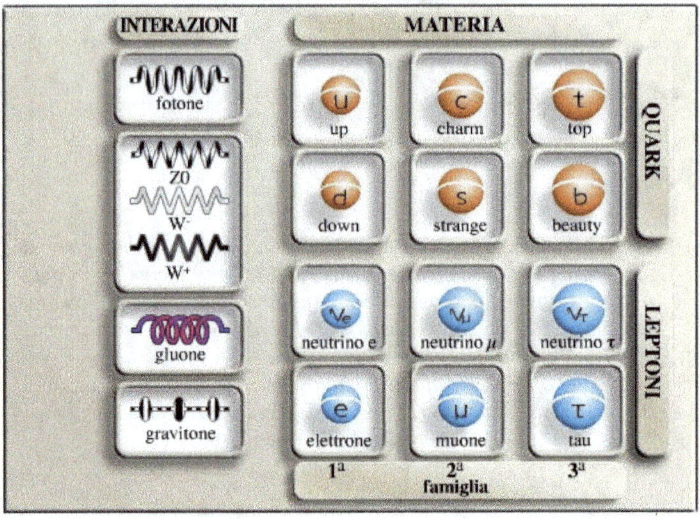

Figure 51

The Standard Model organizes the components of the universe: forces and matter. However, dark energy and dark matter, about which very little is known, are obviously missing. The hypothesis of dark matter, which is thought to constitute the majority of the matter in the universe, comes from experimental observations that indicate significant gravitational effects without corresponding observable matter. No predictions about the nature of such matter can be derived from the Standard Model. In this diagram, unlike in Figure 5, the graviton is also included, even though its existence remains hypothetical. Gravitational waves were detected in 2015, but this does not necessarily imply that the graviton exists. These shortcomings prevent the Standard Model from being considered a complete theory.

As seen in Figure 51, there are four types of forces or interactions: electromagnetic, strong nuclear, weak nuclear, and gravity. Each force is carried by bosons. For decades, researchers have been working to see if it is possible to unify them into a single force, as different aspects of a fundamental interaction. Science sometimes makes prodigious leaps, while in other periods, it advances slowly, step by step. The first step towards a "unified interaction," taken in the 1960s, thanks mainly to the work of Sheldon Glashow, Steven Weinberg, and Abdus Salam, was the development of the "electroweak theory," which jointly describes electromagnetic and weak nuclear interactions. There are technically four fundamental interactions, but two are better than none. Thus, the three scientists shared the Nobel Prize in 1979. However, the electroweak model has a problem: it cannot explain why most of the particles in nature have mass, nor can it explain why they have such different masses. Mass plays a fundamental role in the functioning of the universe. If things had no mass, the universe as we see it would not exist; it would still be a mass of particles moving at the speed of light. However, there is something elusive about the logic behind how nature has endowed particles with mass. Take the electron, an essential component of the atom. There are two seemingly useless "copies" of it: the muon and the tau, which differ only in their mass. And what about the quarks, of which there are six varieties? We have three quarks with a positive electric charge (up, charm, top) and three with a negative electric charge (down, strange, and bottom). If not for their mass, quarks with positive and negative charges would be nearly identical.

Name	Mass in GeV/c^2	Cherge
Up	0,005	+2/3
Down	0,01	-1/3
Strange	0,2	-1/3
Charm	1,5	+2/3
Top	180	+2/3
Bottom	4,7	-1/3

Keep in mind that quark masses are "approximate masses." It is very difficult to determine the mass of a quark, and it is not even easy to define what is meant by "mass of a quark," since a quark can never be isolated. However, enormous differences stand out: the top quark has a mass thirty-six thousand times greater than the up quark, both of which have a positive charge. The same goes for forces. There are clear symmetries that are, however, canceled out by the difference in mass. The strength of the electromagnetic force and the weak force are inherently equal. However, the effects differ greatly because, while the photon, the mediator of the electromagnetic force, travels massless at the speed we know, the two particles that transmit the weak force, the W and Z bosons, are heavier than an iron atom. Their range of action is therefore limited. In fact, they are carriers of the "weak force." So, to understand the deeper reason for these differences, scientists have asked the question: how did mass originate? Usually, when we know how something originated, we understand its function. There shouldn't be anything unnecessary in the universe. If something has come into being, it means it serves a purpose.

In the 1960s, independent groups of theoretical physicists worked to answer this question. The first group had only one member, Peter Higgs, at the University of Edinburgh. The second had two members: François Englert and Robert Brout, in Belgium. The third had three: Gerald Guralnik, C. R. Hagen, and Tom Kibble, at Imperial College London. They all worked on the idea of "spontaneous symmetry breaking," which we discussed in the previous chapter. The three groups were unaware of each other's work until they saw their respective articles published in scientific journals. Then they realized (often with dramatic human implications) that they had been working on similar ideas as others. It was a race against time. Whoever arrives first goes to Stockholm for the Nobel Prize and makes history. They all reached more or less the same conclusions around the same time. It was 1964.

The Glashow/Weinberg/Salam electroweak model was modified. A new field was introduced, later called the "Higgs field," although, as we have seen, many contributed to it. The Higgs field permeates the entire universe. Obviously, a particle is associated with the field: the Higgs boson. The theory states that at extremely high temperatures, immediately following the Big Bang, the Higgs field was symmetric, there was only one interaction, and the W and Z bosons were massless, like the photon. When it began to cool down (so to speak!), the symmetry of the Higgs field spontaneously broke. It "froze," condensed, becoming like a strange, omnipresent gelatin. Particles interact with the Higgs field/gelatin in different ways. The W and Z bosons, getting entangled in the field, absorb mass from it, severely compromising their ability to move. The photon, on the other hand, does not mix with the field, does not get entangled, ignores it, and remains light and fast. Thus, two new types of interactions are created: that of the photon, which is the electromagnetic interaction, operating through a field that moves at the speed commonly defined as "the speed of light," and the weak interaction, which deals with interactions between leptons, between quarks, and between leptons and quarks. The theory formalized by Higgs and others is consistent. It finally explains in a reasonable way the origin of mass, the "why things have weight" from which we started.

But the scientific world is not easily convinced. In the summer of 1965, Guralnik (from the third group) even heard none other than Heisenberg say at a conference in the Bavarian town of Feldafing: "This theory is just garbage." Heisenberg had a notoriously difficult personality, but indeed the problems the theory presents are significant. Among them, the most relevant is that it cannot calculate the mass of the Higgs boson. For a theory on the attribution of mass, being unable to predict the mass of the responsible boson is not a minor flaw. The Nobel Prize seems distant.

Experimental proof would be needed, meaning that the Higgs boson should be "seen." It took about twenty years to find Pauli's neutrino. How long will it take to find the Higgs boson? Assuming it can be found. And assuming, even more importantly, that it exists at all.

XXI. THE FORCES

Let's pause for a moment. The universe is made up of matter particles and force particles. We've seen this. It seems we now need to understand a bit more about what force particles are, since, as we know, they are responsible for inducing interactions between matter particles. There are four interactions that govern the grand dance of the universe. We have seen that the electroweak theory unifies, at least in theory, two of them: the electromagnetic and the weak forces. Forces that operate on very different levels. The electromagnetic force is the one we are all familiar with. It includes not only light, but also the invisible force that allows us to use cell phones and watch satellite television, etc. The weak force plays a mysterious role. It is responsible for some types of radioactive decay. Inside the Sun, the weak force transforms hydrogen into deuterium, a heavier form of hydrogen. Deuterium is the starting point for the nuclear fusion reactions that make the star shine. The weak force does not propagate; it has a range of only one hundredth of a millionth of a nanometer, which is about 1% of the size of an atomic nucleus. In practice, it acts by contact.

The strong force is indeed strong: about 137 times stronger than the electromagnetic force, about a hundred thousand times stronger than the weak force, and 10^39 times stronger than gravity. Its job is to hold particles together: quarks within protons and neutrons; protons and neutrons within the nucleus. And it is by reasoning about the nucleus that we understand why this force is so strong. Protons have a positive electric charge and would repel each other due to the repulsive electromagnetic force. Thus, each nucleus would have a very short life: it would explode. The strong nuclear force serves to counteract the repulsion between protons. The strong force that holds the nucleus together is actually a residual force of the one that binds quarks together. The strong interaction at the quark level occurs with the exchange of gluons, which have a charge called “color charge,” different in nature from electric charge, which continuously changes with the transfer of gluons between quarks, which themselves also possess a color charge. There is no way to separate quarks: the more one tries to separate them, the more the field of forces resists. When we imagine an interaction as a force between two particles, we think of examples such as gravity or electromagnetism, where the force increases as the distance between the particles decreases. But the strong nuclear force does not behave this way; it is characterized by “asymptotic freedom.” When they are “at the asymptotic limit of zero separation,” that is, when they are stuck together, the two particles do not feel any force and are completely free. But if the distance increases beyond the nucleon's boundaries, the strong force tightens its grip and prevents them from moving apart further. It is as if the quarks were attached to the ends of a very strong elastic band. When the quarks are close inside a nucleon, the elastic is slack: it exerts no force. The force/elastic only comes into play when we try to separate the quarks. If the elastic is stretched too much, it breaks, but it does so only at energies sufficient to create quark-antiquark pairs out of the vacuum. It is not possible to extract a quark from a nucleon without creating an antiquark that immediately pairs with it, forming a meson, and another quark that will replace it inside the nucleon. Isolated quarks are impossible to observe.

Gravity was the first interaction force studied by humans. And today, it is one of the greatest obstacles to understanding the universe. It is the weakest force by far. It is ten thousand billion billion billion billion times weaker than the electromagnetic force that binds the electron to the nucleus in a hydrogen atom. That's why the sandwich on the table doesn’t fly towards me. And it wouldn’t fly even if there were no terrestrial gravity. Too little mass involved. Too little to overcome the electromagnetic force that keeps the sandwich there and me here. For gravity to manifest, there must be an enormous amount of atoms all in one place. To attract the sandwich, I would have to be much larger, much larger, an enormous ball of flesh at least five to six hundred kilometers in diameter. The study of gravity is emblematic of the path of scientific thought. For Aristotle, bodies of different weights are attracted to the Earth at different speeds. In the 6th century A.D., a certain John Philoponus, a Byzantine scholar and great commentator on Aristotle, and director of the school of Alexandria in Egypt, contradicted this theory. Discussing the motion of heavy bodies, Philoponus described the experiment that was taken up centuries later by Galileo Galilei: by dropping bodies of different weights together, they reach the ground simultaneously. But in schools, Aristotle's teaching continued to be taught, under the authority of the phrase "ipse dixit," meaning "he said it!" In the teleological Middle Ages, even gravity was explained through God, who gave everything created, living or inanimate, an "instinct" or a nature appropriate to the general order. Dante, in the first canto of Paradise, puts it this way: "In the order I speak of are inclined / all natures, by different fates, / closer to or farther from their origin; // Hence they move to different ports / through the great sea of being, and each one / with an instinct given to it to carry it on. // This carries fire upward to the moon; / this is the driving force in mortal hearts; / this binds the earth and makes it one." So, gravity "binds and makes the Earth one." Dante considered himself a great scientific communicator. And he was considered so by his contemporaries. He was able to transform the driest theological/scientific argument into poetry. That is why I consider this Italian of the 14th century to be one of my teachers.

Even in the Renaissance, schools taught that gravitational acceleration is proportional to weight, despite the fact that the scientists of Alexandria had already demonstrated the contrary. Until the 17th century, the century in which science gradually established itself as experimental knowledge. Newton performed the ultimate act of intelligence by connecting the motion of the stars and planets to the fact that things fall. And he put the "theory of universal gravitation" in writing, accompanying it with extremely complicated calculations made possible by the invention of calculus, also invented by him (Leibniz, in Germany, meanwhile, was inventing it on his own). However, Newton could not explain how gravity occurred, given that there was nothing between one planet and another. How was this force exerted? He got away with talking about "action at a distance," without specifying what it meant. Today we know, thanks to Einstein's general theory of relativity, published in 1915, that gravity is an apparent force. It is not a force, like all the others, but it is the property of matter to deform space-time. That is, gravity is not an interaction at a distance between two masses, but a phenomenon resulting from the deformation of space-time. The presence of mass causes a curvature of the "metric structure" of space-time. If a body moves in "free fall," that is, without being subjected to other forces, it cannot help but follow geodesic paths in space-time. But geodesic paths are not straight lines if space-time is curved. The planets of the Solar System, therefore, have elliptical orbits not due to an attractive force directly exerted by the Sun, but because the mass of the Sun curves space-time. And the Moon thinks it is moving straight, but instead runs in a closed ring-shaped corridor of energy created by the Earth's mass. The gravitational field around a star is represented by the Schwarzschild solution to Einstein's equations, a solution obtained assuming the properties of "spherical symmetry in three-dimensional space independent of time." That same omnipresent symmetry we talked about in a previous chapter. For almost all the planets in the Solar System, the difference between the orbits calculated by Schwarzschild and those described by Kepler's laws (solutions to Newton's equations) is not observable because it is much smaller than the perturbative effects due to the interaction of the planets with each other. But in the case of Mercury, the difference is significant. Newton's gravitational theory could not explain it, Einstein's relativistic theory could. Astonishing!

The explanation of gravity through the curvature of space has fascinating consequences. It changes the concept of falling. Let's imagine firing a projectile with great force. The projectile goes straight, but since space is curved due to the Earth's mass, sooner or later it will hit the ground. However, if the push is strong enough to make it go 28,080 kilometers per hour, by the time space curvature makes it fall, the Earth beneath it will be gone. The projectile will be "in orbit," it will continue to fall without ever hitting the Earth. It will fall forever in a circle. The speed needed to go into orbit is independent of the mass of the projectile and is related exclusively to the mass of the body that curves space. It can be calculated with the Schwarzschild formula. It is called "escape velocity." The greater the mass of the celestial body from which you want to "escape," the greater the escape velocity will be. Incidentally, the escape velocity explains why black holes are black. They have such a large mass that the escape velocity needed to leave them is greater than the speed of light. But nothing can go faster than light. So nothing leaves black holes. Not even light.

So, is everything clear? It would seem so. Experimental evidence of all kinds confirms the correctness of Einstein's theory. We now know that if an apple falls on our head, it is not because some mysterious force is pulling it towards the Earth, but because space around the Earth is curved, etc. But there is a big caveat!

Contemporary physics is dominated by two formidable theories that have explained a myriad of things over the decades, giving us a completely new vision of the immensely large and the immensely small. The great challenge is the unification of the two theories. The dream is to arrive at a single theory: the Grand Unification Theory, GUT. Much progress has been made in this direction, but the conclusion is not yet in sight, not even remotely. And one of the biggest obstacles is gravity itself. All interactions occur through the tireless work of bosons: gluons that hold the atomic nucleus tightly together and photons that form every type of electromagnetic field. Each force has its worker. But who is gravity's worker? We know the others and have seen them at work, but not this one. We have given it a name: "graviton," what else? But we have never seen it. It is partly due to its absence that the Standard Model does not represent a "complete theory of everything." Before 2012, it shared the blame with the Higgs boson. Now it stands alone.

XXII. THE GOD PARTICLE

Here we are. It's time to take a closer look at the main character of this book. We've already mentioned it. Briefly, but we've said important things that now need to be revisited. We've seen that its presence is crucial. Or rather, we've seen how its absence was crucial, an absence that cast shadows over the Standard Model. Let's briefly recall why it is so central. One might wonder: why must there have been a "creator" of masses? Can't we imagine that everything has always been as it is? In the sense that: can't we imagine that each particle was born as it is now, without there having been a precise moment when it acquired mass? Naturally, we are talking about those particles that have mass. There are two answers.

The Big Bang, the primordial explosion of our universe, was solely and exclusively energy. The most accredited hypotheses tell us that at the origin of time and space (we now know they are a single reality), at the zero point of everything, there was a point of zero dimensions and infinite energy. Exactly. Infinite temperature. A heat beyond any possible measure. That point/moment where human imagination is lost. And where experimental science reaches only through simulation. The Big Bang is the extreme boundary of knowledge and imagination. Ultimately, we don't know what happened there. Nothing for certain, I mean. But we do know what happened immediately after. And in this case, "immediately after" truly means immediately after. A hundred billionths of a second after the Big Bang, the universe existed. But it was a very different universe from the one we live in. The initial infinite energy exploded, fragmenting into billions and billions of energy particles, only energy, flying away, distancing themselves from the point of origin and creating space/time. It's logical: when everything was contained in a single point, space/time didn't exist. It didn't exist yet. A hundred billionths of a second later, it did. Those particles that are creating the space/time in which we live had no mass. We have no doubts about this. Today many of them do have mass, so something must have happened. It had to. And that is the first answer.

The second answer is more subtle. And delicate. It's the "poetry of logic" Einstein wrote about. The one that creates ice crystals and the equations that describe them. It tells us that there is an understanding between human intelligence and nature. It tells us we are on the right track. It tells us "if" we are on the right track. We've talked about it before: it's symmetry. The queen of forms. And her "wicked" sister: the breaking of symmetry. A queen too. Queen of transformation. Wicked in the sense that she breaks what is perfect. But she breaks it so that new perfections can come into being. The primitive universe is perfectly symmetrical. A kind of fog of equal density at every point. All particles are the same. All without mass. Then something happens. A break in the initial symmetry. Because the particles, moving away from each other in their frenzied race at the speed of light, decrease the overall energy density. The temperature drops. The cooling process begins, the fate of everything that expands in space.

Another consideration. Remember what we said about the photon? For it, space/time practically doesn't exist. It's all speed. A particle of pure speed is like a point moving on a line. It has no body, it ignores three-dimensional space. There is no other solution; energy is not enough. For a three-dimensional universe (oops, four-dimensional) to be created, mass is absolutely needed, that is, new physical realities capable of moving in the three dimensions of space and in the dimension of time. A part of the force particles needs to slow down, to condense a percentage of their energy into matter. How can this happen? They can't do it on their own. A field is needed that fills the entire universe and breaks the initial symmetry, giving rise to a new story.

If you want, the question, "Why is it so important to know the

origin of mass?" could be answered in other ways, more difficult to explain in simple words. There are concepts that are difficult in themselves because they describe realities particularly strange, very far from our experience. Let's try.
In the theoretical formulation of electroweak symmetry (which, let's remember, says that once upon a time electromagnetism and the weak force were one and the same), both the photon (mediator of electromagnetism) and the weak vector bosons (mediators of the weak force) initially have zero mass. So, we are in the presence of symmetry. Electroweak symmetry. Then the large masses observed experimentally for the weak vector bosons originate from the breaking of electroweak symmetry. How did this happen?
Here's the first challenge to our way of thinking. We must strive to imagine a vacuum that actually contains something. The Higgs field exists in this "non-zero vacuum." The vacuum, therefore, does not coincide with our idea of "nothingness." At least the quantum vacuum doesn't. It is now established that the quantum vacuum is in constant fluctuation. In this non-zero vacuum, there is the Higgs field, which is a "uniform non-zero field" capable of giving mass to particles that lack it.
Difficult, isn't it? But let's continue.
Now let's clarify better what is meant in particle physics by "mass." Normally, this term is used as an abbreviation for the more precise term "rest mass," which has a slightly more restrictive meaning than the concept of inertia in classical physics. Every moving particle has a certain energy, and this energy is always associated with a certain amount of momentum, and therefore a tendency to resist changes in the state of motion, that is, a certain inertia. In some particles, when the momentum tends to zero, their energy also tends to zero. These are the massless particles. For the photon, in fact, the "rest mass" is zero. For other particles, however, the "rest mass" retains a certain value even when the momentum is zero. It is the intrinsic energy, residing in the mass. Let's remember Einstein's famous formula, which tells us that mass and energy are expressions of the same reality and that mass can become

energy and vice versa. Now, the fact that some rest masses are zero while others are not contradicts the principle of symmetry, in one of its specific versions: "local gauge symmetry" or "local gauge invariance." This principle is the foundation of the theory behind the Standard Model, whose success over the decades in explaining a vast number of phenomena has been extraordinary. Physicists value the Standard Model more than their right foot and cannot admit to contradicting it by attributing mass to some particles and not to others. If it were admitted that massive particles have "intrinsic mass," it would contradict local gauge invariance. That is, it would render inconsistent the entire theory that allows us to fundamentally understand the interactions between particles. So, it must be assumed that all particles have "zero intrinsic mass." Only, as we have already seen, some of them perceive the presence of a field, couple with it, and acquire a "non-zero rest energy." It's the Higgs field, we now understand. Which, at a fundamental level, respects gauge symmetry and therefore allows us to simultaneously explain both the fundamental interactions between particles and the presence of non-zero rest masses. As Jim Baggott wrote, "The concept of mass has evaporated in a puff of logic, replaced by interactions between the Higgs field and otherwise massless particles."

We can try to further clarify the idea by considering the behavior of the electron when it passes through a solid. A solid is a crystalline lattice of atoms. When the solid is crossed by an electron, the atoms in the lattice are attracted to it. Consequently, the mass of the electron increases up to forty times that of a free electron. Similarly, to explain the mass of particles, the idea was conceived of a background field that thickens locally when crossed by a moving particle. This thickening of the field around the particle generates the particle's mass. The Higgs field in the vacuum can thus be imagined as a crystal filling our universe.

Furthermore, we are capable of constructing "Faraday cages," which are boxes within which electromagnetic fields cannot enter. It is not, for now, technologically conceivable to

construct "Higgs cages." The Higgs field is defined as "unshieldable." But imagining Higgs cages helps to understand what the Higgs field is. If there were a Higgs cage, everything put inside it would immediately disappear, turning into energy. All particles would go back to running free at the speed of light. If it were possible, humanity would have solved in one fell swoop the problem of waste and that of energy. However, it would have to face the problem of controlling that energy, which would be truly immense.
Capite queste cose, non è difficile capire perché il bosone di Higgs ha cominciato a essere chiamato con il lusinghiero epiteto "particella di Dio". Dando massa alle particelle, è lui che ha permesso, e permette ancora, alle cose di esistere. Il bosone di Higgs "crea" la materia, con la quale l'universo diventa concreto. Le particelle/materia, aggregandosi, danno origine agli atomi, alle molecole, agli elementi, ai mondi. E anche a noi.

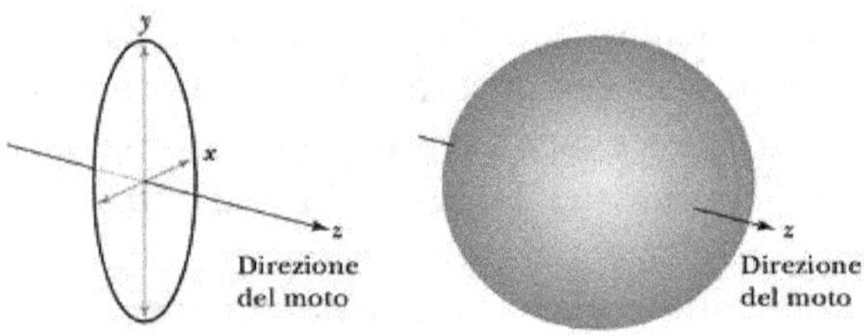

Figure 52

There are many ways to explain the "God particle." In this diagram, the matter of its origin is explained through the attribution of the "third degree of freedom." The particle before Higgs, massless, has only two degrees of freedom. Its oscillation (let's remember that every particle is also a wave) can only occur along the y and x axes. By interacting with the Higgs field, it acquires a third degree of freedom: it now oscillates also in the z direction of motion. It has become three-dimensional.

In the early 1960s, Higgs and the others were frantically sifting through formulas. We left them, in a previous chapter (XIX), grappling with Nambu's equations.
During those years, many things were clear. Physicists had worked hard, and the landscape appeared well-defined in every detail. In summary:

1. All matter is made of chemical elements.
2. Chemical elements exist in a great variety of possible forms, represented in Mendeleev's periodic table: from the lightest, hydrogen, to the heaviest naturally occurring element, uranium.
3. Each element consists of atoms.
4. Each atom consists of a nucleus made up of a variable number of positively charged protons and neutrons.
5. Each element is characterized by the number of protons in the nucleus of its atoms. Hydrogen has one, helium has two, lithium has three, and so on up to uranium, which has 92.
6. Electrons, which are negatively charged and in a number sufficient to balance the number of protons and make the atom globally neutral, are found around the nucleus.
7. Each electron can have a spin orientation either up or down, and each orbital can hold up to two electrons, provided their spins are opposite.

This set of certainties, with protons, neutrons, and electrons as fundamental building blocks, combined with Pauli's exclusion principle, explained why matter has form and density.
Quarks would arrive a few years later.
We have also seen that the Higgs boson addresses a vital necessity: the Standard Model needs it in order not to collapse, sweeping away decades of research. But according to most physicists of the time, the real problem of the origin of mass had been solved by Higgs and company with a smoke and mirrors act, a series of tricks: a hypothetical field, a hypothetical boson. Better to focus on something else.

XXIII. QUARKS AND THE UNIVERSE

Interest in the Higgs boson grew in the 1970s and 1980s when results from particle accelerators experimentally confirmed the validity of electroweak theory. Physicists had learned from experience that electroweak unification could now be applied to a broader problem. Electroweak theory implied that, moments after the Big Bang, at the extremely high temperature of the universe, the weak nuclear force and the electromagnetic force were indistinguishable, merged into a single electroweak force carried by massless bosons. This is the "electroweak era" of the universe. As the universe cooled, the Higgs background field, as we have already seen, "crystallized," and the gauge symmetry of the electroweak force broke. The massless bosons of electromagnetism, the photons, continued to remain free, while the bosons of the weak force, interacting with the Higgs field, acquired mass and transformed into the W and Z particles (theorized by Glashow, Weinberg, and Salam in 1968 and discovered with CERN's superaccelerator by Carlo Rubbia in 1983). This is why the weak and electromagnetic forces appear separate and so different today.

In those years, physicists discovered astonishing things. In 1974, Weinberg, Howard Georgi, and Helen Quinn demonstrated that the intensities of the three fundamental forces among particles become nearly equal at energies between one hundred billion and one hundred million billion GeV (giga electron-volts). Pay attention to the numbers: when the universe was one hundred billionth of a billionth of a billionth of a billionth (10^-35) of a second old, it had a temperature of about ten billion billion billion (10^28) degrees. It then seemed reasonable to assume that, during that era—the era of the "Grand Unification"—the strong nuclear force and the electroweak force were also indistinguishable and merged into a single "electronuclear" force. All the force mediators were identical, and there was no mass, electric charge, flavor (up, down, etc.), or color (red, green, blue) for quarks. Breaking this symmetry required other Higgs fields, which crystallized at higher temperatures, causing a division between quarks, electrons, and neutrinos and between the weak and strong forces.

The numbers we just saw—astonishing values of time and temperature—are the result of precise and irrefutable calculations. But no terrestrial collider will ever be able to reach that temperature. Does this mean that the GUT, Grand Unification Theory, is unverifiable? It was necessary to circumvent the problem, to reach it by other, collateral means. For example, reasoning about the particles necessarily created at those temperatures. And then, even though that era is billions of years in the past, it has left indelible marks on our universe.

In the late 1970s, cosmologists began to take an interest in Higgs fields. Alan Guth and Henry Tye introduced the concept of "supercooling" linked to an extraordinary expansion of space-time. Only in this way could the disappearance of certain particles, considered certain during the grand unification period and now completely nonexistent—"monopoles," that is, tiny magnets with only one pole—be explained. Initially, Guth himself was puzzled by the result. But he soon realized that this explosive expansion managed to explain important characteristics of the observable universe. Things that the then-popular Big Bang cosmology could not explain. This is the theory now embraced by many called "inflation." In this context, Higgs fields are called "inflaton fields." Inflation explains many things and answers questions that the Big Bang alone cannot answer. Today, it is more than just a hypothesis, having been confirmed by several experimental results. Among these were the spectacular results of April 1992 when the COBE (Cosmic Background Explorer) satellite produced a map of the tiny temperature variations of the cosmic background radiation: the cold remnant of the radiation that separated from matter about four hundred thousand years after the Big Bang.

To summarize: Brout, Englert, Higgs, Guralnik, Hagen, and Kibble invented the Higgs field as a tool to break the symmetries at the origin of the universe. Weinberg and Salam applied the same mechanism to electroweak symmetry breaking and to correctly predict the mass of the W and Z bosons. The breaking of electronuclear force symmetry, finally, led to the discovery of inflationary cosmology and the precise prediction of the large-scale structure of the universe. A completely theoretical concept has, over time, had unforeseen and surprising implications, becoming central both to the Standard Model of particle physics and to the new model of cosmology. Now there was only one question left. Everyone was asking it: do these extraordinary Higgs fields really exist? There was only one way to find out.

XXIV. NOW IT NEEDS TO BE FOUND

As we have seen, in the decades following its theoretical formalization, the Higgs boson gained increasing importance in everyone's eyes. But physics needs evidence. Proof, for a particle, is to find it, to see it. And this can only happen in extreme conditions, artificially created in the laboratory. CERN's LHC in Geneva is the largest laboratory ever built. But not the first.
When the Second World War broke out, the United States faced the challenge of obtaining sufficient quantities of uranium-235 for the construction of the atomic bomb, which would later be dropped on Hiroshima. It was a race against time. There was a fear that Hitler's Germany might be the first to build the terrible weapon. At Oak Ridge, Tennessee, the Y-12 plant was set up for the electromagnetic separation of isotopes, based on the cyclotron developed in the 1930s by Ernest Orlando Lawrence, a heavy charged particle accelerator. But the one from 1943 had titanic dimensions: it used magnets 76 meters long that weighed between three thousand and ten thousand tons. All the copper reserves of the USA were used to build them. The Treasury lent the Manhattan Project fifteen thousand tons of silver for the coils. The magnets consumed as much electrical power as a large city. The plant, which began operating in November 1943, employed thirteen thousand people. It was the first example of "big science." It took a dramatic urgency, a world war, for a great industrial country to decide to employ so many resources in a scientific endeavor.
At the end of the conflict, the physicists who had created the atomic bomb and enabled the United States to win the war became the darlings of Congress and managed to secure government funding for increasingly costly experimental machines. A kind of grand race began between the USA and Europe, with alternating results. In 1983, the Tevatron accelerator

was launched at Fermilab in the USA. In twelve hours, the beam inside the six-kilometer ring reached the design energy of 512 GeV (giga electronvolts). By colliding protons and antiprotons, the energy would be 1 TeV (tera electronvolt). It had cost 120 million dollars. But it wasn't enough. Higgs bosons are very heavy particles and require enormous energies to be created. A new project was launched, with a projected cost of four billion dollars: the Superconducting SuperCollider (SSC). Congress wavered. Many thought it was too much. They couldn't spend all that money just to "please" a few thousand physicists, even if they had won the war. The physicists in charge of the project were continuously summoned to Washington to explain and re-explain the purpose of the operation. Then came the moment to allocate the funds. In the decisive session, at the end of the debate, all eyes were on Ronald Reagan, the president. It was up to him to have the final say. It was January 1987. Other scientific projects, of lesser scope but with more tangible practical results, were knocking on the door. Reagan took the floor and quoted Jack London, who cited a phrase from the famous American football player Ken Stabler, known as Snake: "When you're in trouble, throw deep!" The project was approved.

But it would face a tough life. Deciding where to build it was a long and arduous process. Every state wanted that shower of billions. In the end, Texas prevailed. Meanwhile, the projected costs had doubled. President George Bush, Reagan's successor, kept the project going but sought international support. He pushed Japan, which was very wealthy. Japan refused: we don't have to pay for America's anxiety over scientific glory. In 1990, digging began in Texas. Laboratories were built for the development and testing of magnets. Facilities were constructed for the enormous "refrigerators" needed to produce and circulate the liquid helium required to keep the magnets at extremely low temperatures, and so on. Two thousand scientists from all over the world were already engaged. But shadows were gathering. There was resentment among scientists. All that money to find the Higgs boson! And what about other projects? How many could be funded with all those billions! The expression "big science" risked becoming synonymous with the waste of public funds. That's how taxpayers' money is wasted! Just the detector threatened to cost 500 million dollars! Congress was increasingly uneasy. In 1992, Bill Clinton became president, immediately less determined than Reagan and Bush in defending the SSC. A new program came into play: 25 billion dollars for the construction of the International Space Station in Houston, also in Texas. In October 1993, after the last desperate attempts by Weinberg, Richter, and Lederman to save the project, Congress approved the International Space Station. The day after, it cut funding for the SSC. Writer Herman Wouk would later publish a book about the case, titled *A Hole in Texas*.

Figure 53

There are still 23 kilometers of abandoned tunnels in Texas, excavated for the SSC project.

XXV. JULY 4, 2012

The failure of the SSC is symptomatic of a deeper issue. It is essentially a communication problem. Physicists now have the task of helping ordinary people, and politicians, understand what they are doing. This is not easy. But since these kinds of research projects require laboratories with staggering costs, they cannot avoid it. They must be convincing not only in highlighting the scientific value of their research but also, even more importantly, in predicting appealing technological benefits and practical advantages. It is not enough to say that without interactions between fundamental particles and the Higgs field, there would be no matter, no planets, no stars, no life. Ordinary people have a strong desire to live their lives well, and knowing the origin of life is not strictly necessary for living it well.

One year after the American withdrawal, Europe took up the challenge. On December 16, 1994, CERN member states voted to allocate 15 billion dollars over twenty years to transform the Large Electron-Positron Collider (LEP) into the Large Hadron Collider (LHC), a proton-proton collider. The new facility would be capable of providing collision energies of 14 TeV, less than half the maximum energies planned for the American SSC, but sufficient to discover the Higgs.

Meanwhile, other very interesting developments were occurring in the world of physics, such as "supersymmetry," which clarifies many of the dark corners of the Standard Model. Essentially, supersymmetry, affectionately called SUSY (from Super Symmetry), says that every particle is paired with a partner particle. SUSY was developed in the 1970s by Jonathan Segal and his group at MIT, simultaneously with Daniel Laufferty at Tufts University, and two Soviet theoretical physicists, Izrail' Moiseyevich Gel'fand and Evgeny Pinkhasovich Likhtman. It was born in the context of string theories: the vibrational modes of strings that give rise to fermions and bosons necessarily come in pairs. For now, the existence of supersymmetry is based only on theoretical considerations. Supersymmetric particles have not yet been observed. If supersymmetry exists in the real physical world, it must necessarily be a broken symmetry. Only in this way can it be explained why these particles have not yet been observed: they are too heavy. Extraordinary energies are required to produce them. It is hoped that the Large Hadron Collider will be able to confirm or refute the theory.

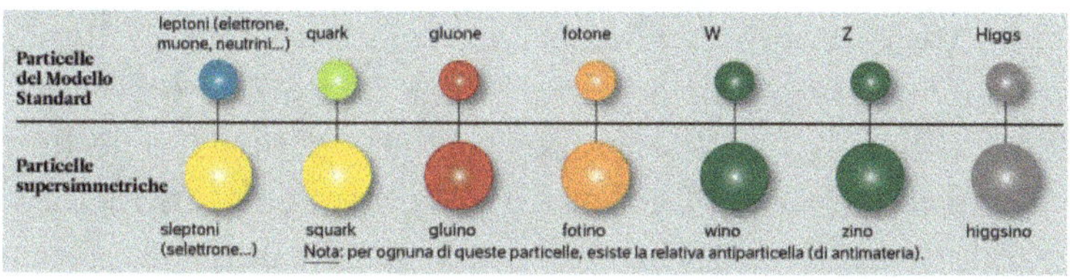

Figure 54

SUSY, the supersymmetric theory, predicts that each particle has a partner: an "antiparticle." The mass of the antiparticle is greater than that of the corresponding particle due to the breaking of symmetry. The fact that they have a large mass would explain why they have not yet been observed.

In 2008, the LHC was ready to be turned on. It has gigantic proportions. The construction took eight years. There are 1,746 superconducting magnets kept at 271.4 degrees below zero, the lowest temperature in the universe. To bring the magnets to this temperature, ten thousand tons of liquid nitrogen and 150 tons of liquid helium were used in a circuit. There are four detectors. One of them, ATLAS, is 25 meters high, 48 meters long, and weighs seven thousand tons. And so on. After an incident with the cooling system, which halted everything for about a year, operations resumed in November 2009. And the parade of already known particles began, with the Standard Model in full display. Meanwhile, rumors of 3 sigma signals of a "light" Higgs were coming from the American Fermilab.

(The "sigma" is a complicated system for calculating the level of certainty. 3 sigma corresponds to a certainty level of 99.7%. It's not enough to claim a discovery. Physicists require a 5 sigma, or 99.9999%, which means a 0.0001% chance that it's just a coincidence.)

Exciting but unconfirmed news was also leaking from the LHC. Physics blogs around the world were buzzing; it seemed like something big was about to be announced. Inside the ring, protons were racing at speeds very close to that of light, one beam in one direction and the other in the opposite. At that speed, millions of collisions per second occurred. The luminosity continued to rise. On May 3, 2011, a peak luminosity corresponding to 75 million collisions per second was recorded. Calculations showed that this meant 0.011 Higgs bosons per second, or one Higgs boson on average every minute and a half. But it's important to note that "seeing" a particle doesn't mean seeing it with your eyes or simply photographing it as you would something in the macroscopic world, or similar things. In this case, seeing means that the detectors, like gigantic cameras, take millions of shots per second. What do they photograph? Light trails. The Higgs boson has a short lifespan and can decay in different ways, each leaving a distinct signature in the detectors. For example, sometimes the Higgs disintegrates in a flash of two high-energy photons. Other times, in a jet of four leptons—electrons, muons, or their heavy relatives. The trails left by these decay particles provide most of the information. Many of these "photographs" are immediately discarded. This means that the data they contain are not considered. These are the ones involving already known particles and are therefore useless. The others must be studied in every detail to understand if they are indeed the result of new particles. Therefore, mountains of data collected during the collisions must be assembled, subjected to meticulous analysis, calculations performed, and conclusions drawn. This involves a lot of time and the coordinated work of thousands of scientists. And thousands of computers scattered around the world. Each computer offers its computing power to achieve a sufficient overall capacity. The transition from the pure and simple physical collision to a point on a graph on a computer screen requires a chain of events involving hundreds of people with various specialized skills.

Throughout the first half of 2011, the data progressively available were encouraging enough that those involved decided to form a small working group so that information from the two large detectors, ATLAS and CMS, could be quickly compared. Meanwhile, there were toasts to the initial successes. Adjusting the collider is incredibly difficult. There are a multitude of variables to control. One example: protons circulate in the LHC at about eleven thousand revolutions per second (each revolution is twenty-seven kilometers), oscillating on two orthogonal planes. If the number of oscillations per revolution is an integer, and there is an integer number of oscillations in half a revolution, in a third of a revolution, or in three quarters of a revolution, and so on, the oscillations amplify over time. If the resonances cannot be controlled, the beams are lost as well. Does this remind you of anything? Even electrons around the nucleus must be able to oscillate an integer number of times. The risk that something goes wrong is always present. The following is an example of a serious danger that was fortunately averted by precautionary measures. Before particles are injected into the main LHC collider, they are accelerated to the proper speed inside an accelerator called the Super Proton Synchrotron (SPS). The moment when the particles transition from the SPS to the LHC is delicate: every part of the device must work perfectly. As soon as the particles enter the LHC, the superconducting magnets turn on, producing instant fields that curve the beam and project the particles in the correct direction. Left alone, the particles would travel straight. And the collider is round. The moment the magnets are activated, the proton beam moves at a speed close to that of light and carries an amount of energy sufficient to pierce almost any metal to a depth of one meter. At 4:47 in the morning of April 15, just as a beam of 108 proton bunches entered the LHC, one of the magnets short-circuited and its field suddenly collapsed to zero. Only 32 proton bunches of the beam entered the LHC following the correct trajectory. The other bunches, not having been deflected, continued in a straight line. If this eventuality had not been anticipated, the beam would have created a hole in the collider. But right on the path of the undiverted beam, a protective absorber plate had been placed.

On July 30, a luminosity of one hundred and twenty million proton/proton collisions per second was reached. On August 7, even more. But all this was still at 3 sigma. Higgs, retired since 1996 in Edinburgh, followed the events with the calm of an octogenarian. "It would be nice, after all this time, to have proof that I was right."

But more time and patience were needed. One hundred and forty thousand billion proton/proton collisions were still not enough to go beyond 3 sigma. On October 31, the number of collisions rose to three hundred and fifty thousand billion. An ocean of data. To store them on regular DVDs, it would take a tower twenty-six kilometers high. The experiment leaders were very cautious. Making a bad impression by being carried away by enthusiasm is always a risk. As happened in February 2012, when news came from the Gran Sasso Laboratory that neutrinos had been intercepted traveling faster than the speed of light. If true, it would have overturned Einstein's special relativity. On February 22, the retraction: a defect in the connection of a fiber optic cable caused a slight delay in the measurement: 73 millionths of a second. A small amount, but decisive. Embarrassment and resignations. The line between glory and shame is so thin!

Figure 55

ATLAS is one of the two general-purpose detectors of the Large Hadron Collider (LHC). It investigates a wide range of physical phenomena and has been involved in the search for the Higgs boson. Although it shares the same scientific goals as the CMS experiment, it uses different technical solutions and a different design for its magnetic systems..

After a few months of maintenance, the LHC restarted on March 12, 2012. On July 4, CERN organized a seminar "to inform the scientific community about the progress of the work." There was an atmosphere of great anticipation. Even the “inventors” of Higgs—Higgs, Englert, Guralnik, Hagen, and Kibble—were invited. This definitely meant something! In the morning, security had difficulty letting groups of ten people at a time into the conference hall. Every seat had to be occupied in an orderly manner. The turnout was huge. Many young scholars had spent the night in sleeping bags, on stair landings, to make sure they could get in. The seminar was streamed worldwide. Some people, once seated, took advantage of the wait to catch some more sleep. Then a murmur and flashes from photographers. Peter Higgs entered, greeted with applause. The young people woke up. “Ah, Peter Higgs, there he is! Fantastic!” The older ones gathered around to greet him. The young ones didn’t dare. They clapped their hands. Higgs sat among the others. Rolf-Dieter Heuer, CERN's Director General, gave a brief welcome: “Good morning, everyone, welcome. Today is a special day. We will present the results of the two experiments, ATLAS and CMS. We won’t follow alphabetical order, and I invite Joe Incandela, CMS Director, to speak.” Joe Incandela, a physics professor at the University of California and head of the CMS collaboration, spoke at length about the enormous complexity of the research and the scientific and technical contributions of all collaborators. He showed slides. “This is a slide for the theorists. And they deserve nothing more. Ah ha, I’m joking! This is the Standard Model. That’s it, this is what we know. [...] But this year, as many of you know, we have increased the values. We worked with much more intense beams and observed many interactions, and it was in one of these events that we were able to see one of the rarest particles ever; the Higgs boson. These are last week’s data. As you know, by observing the radiative corrections of W which are linked to the Top, so knowing the mass of W and the Top mass very well, one can predict a long range. [...] The first significant result is the evidence of Higgs particles decaying into two photons.” The slide showed a red line connecting black dots, outlining a peak in the data. At this point, Incandela stopped talking, stepped back, and stared, as if spellbound, at the slide. He recovered after about ten seconds: “I lost myself for a moment, sorry!” The slides slowly reconstructed the results, step by step. Finally: “Overall, the cleanest channels, those in which the Higgs particles decayed either into two photons or four leptons, give a solid five-sigma signal. [...] So, to summarize, we have observed the new boson with a mass of 125.3 plus or minus 0.6 GeV, with a significance level of 4.9 sigma.” Applause.

Immediately afterward, Fabiola Gianotti, ATLAS Director, took the floor. She had spent the last few nights organizing her presentation for this morning. She was beaming, in a red sweater. She kept adjusting her long black hair. She also tried to appear objective and control her emotions. Meanwhile, as Fabiola Gianotti spoke, trying to appear detached, down below, the ATLAS detector continued to collect data. 1,347 proton bunches were colliding 20 million times per second, recreating the primordial universe. "Good morning, ATLAS is truly pleased to present here and today the updated results of the Standard Model Higgs searches based on data up to 10.7 inverse femtobarns recorded in 2011 and 2012, and it is a great honor and a great emotion for me to represent this fantastic collaboration on this occasion. Let's move on to the results for this channel. Here you see the results for 2011, those for 2012, and the combination of the two [...] So, it's not a theoretical belief, but the sidebands of a possible signal [...] And here's the combination. This distribution is extremely clear, except for a large peak here, in this region. A peak with a significance of 5.0 sigma and a mass of 126.5 GeV." Great applause. Everyone looked towards Higgs, who took off his glasses to wipe his eyes. Heuer took the microphone: "As a layman, I would say: we've found it!" Applause and cheers of joy. Fabiola: "Is Peter here? Is Peter here?" "He's here, Peter is here!" Everyone stood up, continuing to applaud. "Peter, Peter, come, come down." The microphone to Peter: "I want to congratulate again all those who contributed to this great discovery. It is truly an incredible moment in my life." Heuer: "Not just yours, Peter!" Laughter. Applause and laughter. Heuer: "And I think all of us, all the people watching us from various conference rooms, everyone who has been involved in this project can be happy on this day. So, let's enjoy it!"

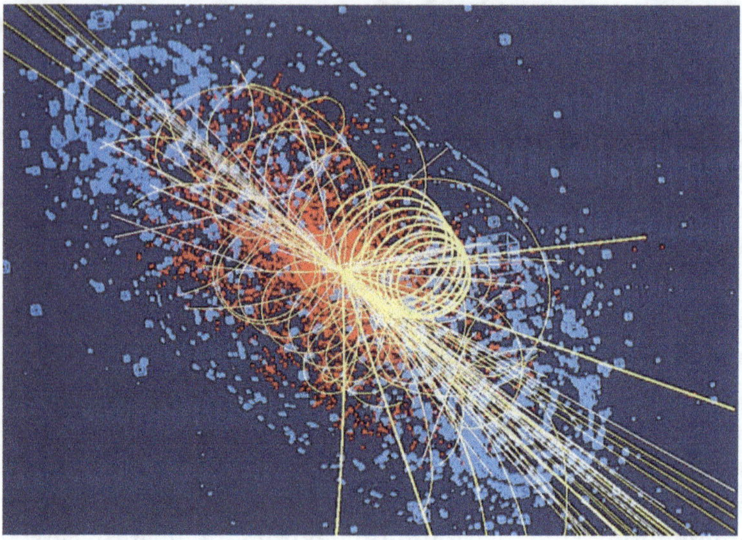

Figure 56

In this CMS simulation, protons collide at 14 TeV, producing four muons. The lines denote other particles.

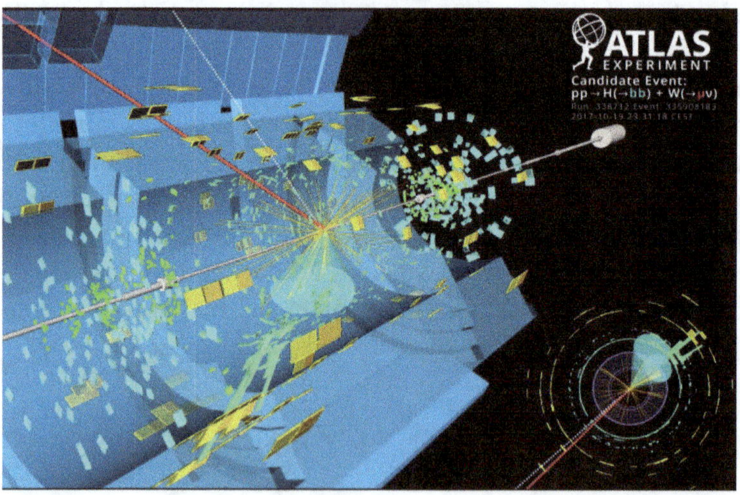

Figure 57

A candidate ATLAS event for the Higgs boson (H) decaying into two bottom quarks (b), in association with a W boson, which in turn decays into a muon (μ) and a neutrino (ν). (Image: ATLAS / CERN).

XXVI. THE "HIGGS"

The discovery of the Higgs boson marks the end of a great era in human knowledge. We now know much about the structure of the universe and its origin. But above all, one thing is now definitively certain: mass does not exist. There is no such thing as "intrinsic mass". Things do not weigh. They simply move more or less fast. The word "mass", so prevalent in physics textbooks, is now definitively changing its meaning. Einstein had said that mass and energy are interchangeable: one transforms into the other and vice versa. Now we know more. Mass is exclusively the energy of the interactions that occur between elementary quantum fields and their particles. The mass of particles is their effort to move. And the field, or fields, that force particles to squeeze their energy into "mass" is the Higgs field.

It will take many more years of study and experiments to understand how many Higgs bosons there are. One, as the Standard Model suggests, or even five as supersymmetry predicts? Most physicists hope that the particle found turns out to be one of the types predicted by supersymmetry, the theory that predicts a heavy companion for every known particle. A supersymmetric Higgs would open the way to a new dimension for physics and take a giant leap towards a theory that explains all particles and all forces of nature. There are still open questions. The theory does not explain the differences in mass. The top quark, which is the heaviest known particle, is about four hundred thousand times heavier than the electron. Why?

We will need further confirmations and more information. But the existence of the Higgs boson is beyond doubt. It took almost fifty years to find it. It took the efforts of thousands of scientists and the financial involvement of states. All the newspapers are talking about it. A new era begins. But I like to end this little book with the human touch that the presence of the Higgs gave to the seminar of July 4, 2012: "It is truly wonderful that it happened while I am still alive."